Titles in This Series

24 **Byron Goldstein and Carla Wofsy,** Cell biology, 1994
23 **Robert H. Gardner,** Predicting spatial effects in ecological systems, 1993
22 **Marc Mangel, Editor,** Sex allocation and sex change: Experiments and models, 1990
21 **Hans G. Othmer, Editor,** The dynamics of excitable, 1989
20 **Alan Hastings, Editor,** Models in population biology, 1989
19 **Gail A. Carpenter, Editor,** Circadian rhythms, 1987
18 **Louis J. Gross and Robert M. Miura, Editors,** Plant biology, 1986
17 **Robert M. Miura, Editor,** DNA sequence analysis, 1986
16 **Robert M. Miura, Editor,** Muscle physiology, 1986
15 **Robert M. Miura, Editor,** Neurobiology, 1982
14 **Stephen Childress, Editor,** Some mathematical questions in biology, 1981
13 **George F. Oster, Editor,** Some mathematical questions in biology, 1980
12 **Simon A. Levin, Editor,** Some mathematical questions in biology, 1979
11 **Simon A. Levin, Editor,** Some mathematical questions in biology. X, 1979
10 **Simon A. Levin, Editor,** Some mathematical questions in biology. IX, 1978
9 **Simon A. Levin, Editor,** Some mathematical questions in biology. VIII, 1977
8 **Simon A. Levin, Editor,** Some mathematical questions in biology. VII, 1976
7 **Simon A. Levin, Editor,** Some mathematical questions in biology. VI, 1974
6 **J. D. Cowan, Editor,** Some mathematical questions in biology. V, 1974
5 **J. D. Cowan, Editor,** Some mathematical questions in biology. IV, 1973
4 **J. D. Cowan, Editor,** Some mathematical questions in biology. III, 1972
3 **J. D. Cowan, Editor,** Some mathematical questions in biology. II, 1972
2 **M. Gerstenhaber, Editor,** Some mathematical questions in biology. I, 1970
1 **M. Gerstenhaber, Editor,** Some mathematical problems in biology, 1968

Cell Biology

Lectures on Mathematics in the LIFE SCIENCES

Volume 24

Cell Biology

1992 Symposium on
Some Mathematical Questions in Biology
November 15–19, 1992
Denver, Colorado

Byron Goldstein
Carla Wofsy
Editors

American Mathematical Society
Providence, Rhode Island

Proceedings of the 1992 Seminar on Some Mathematical Questions in Biology, held at the Annual Meeting of the American Society for Cell Biology in Denver, Colorado, November 15–19, 1992. This symposium was sponsored by the National Science Foundation under Grant Number DMS-9105171 01.

1991 *Mathematics Subject Classification*. Primary 92–02;
Secondary 92C05, 92C10, 92C45.

Library of Congress Cataloging-in-Publication Data

Seminar on Some Mathematical Questions in Biology (1992 : Denver, Colo.)
Some mathematical questions in biology—cell biology : 1992 Symposium on Some Mathematical Questions in Biology : Cell Biology, held November 15–19, 1992, Colorado Cenvention Center, Denver, CO./Byron Goldstein & Carla Wofsy, editors.
p. cm. — (Lectures on mathematics in the life sciences; v. 24)
Includes bibliographical references.
ISBN 0-8218-1175-4
1. Cytology—Mathematical models—Congresses. I. Goldstein, Byron, 1939– . II. Wofsy, Carla, 1944– . III. Title. IV. Series.
QH573.S94 1992 93-43588
574.87′015118—dc20 CIP

∞ The paper used in this book is acid-free and falls within the guidelines
established to ensure permanence and durability.
♻ Printed on recycled paper.
Most of the articles in this volume were printed
directly from copy prepared by the authors.
Some articles were prepared using AMS-TeX,
the American Mathematical Society's TeX macro system.

10 9 8 7 6 5 4 3 2 1 99 98 97 96 95 94

CONTENTS

PREFACE

The molecular biology revolution has changed profoundly the way cell biology is done. Model cellular systems can now be designed to study particular mechanisms. It is almost routine (or at least it seems so to those who don't have to do the work) to create a cell line that expresses, or is missing, one or more proteins whose function the investigator wants to study. With the ability to clone and sequence, and then compare sequences through the use of data bases, the number of known proteins has risen dramatically. This in turn has led to the identification of new families of proteins, bringing considerable order to what at first seemed like a chaotic endeavor. Through recombinant DNA technology, copious amounts of purified protein are obtainable. Researchers now carry out detailed structural and functional studies on almost any protein they wish. Publication of the three-dimensional structure of a protein has gone from being a rare event to a weekly occurrence.

It is not surprising that even some mathematical modelers have been touched by this revolution. Simple, well defined experimental systems, which now abound in cell biology, lend themselves to mathematical description and to the acquisition of quantitative data. (Simple, of course, is a relative term. None of these model cellular systems is really simple, but they are considerably less complex than the model systems of ten years ago. Much heterogeneity has been eliminated, and the systems can be designed to yield answers to specific questions.) Increasingly, mathematical models are required to guide experimental design and analyze the quantitative data obtained. As quantitative methods continue to improve and their application to model systems expands, we expect mathematics to become a widely used and widely accepted tool in cell biology.

In the work described in this volume, theory and experiment combine to answer questions in cell biology. Michael Reed and Jacob Blum clarify the mechanisms of protein transport in nerve axons, solving interesting mathematical problems in the course of the analysis. Eric Odell and George Oster show how the coupling of membrane protein shape and local membrane curvature can account for experimental observations regarding protein localization and vesicle budding. Elliot Elson and Hong Qian review experimental methods for studying the motion of cell membrane proteins and present the theory relating the experimental measurements with the underlying molecular transport equations. Micah Dembo uses a mechanochemical model to classify bonds formed between cell membrane adhesion molecules and sites on a substrate, to interpret data on cell attachment and cell peeling, and to devise a new experiment to distinguish alternative models. Jennifer Linderman, Nancy Berry, and Debra Singer study activation of helper T lymphocytes by antigen presenting cells using a sequence of models and experiments to find quantitative relationships between antigen exposure and antigen presentation, antigen presentation and T cell binding to antigen presenting cells, and, finally, T cell binding and the T cell response. Byron Goldstein and Carla Wofsy use chemical kinetic and geometric models to show how the aggregation of cell surface receptors affects the interpretation of binding and cellular response data in several model experimental systems.

The six papers in this volume expand on invited lectures presented at the 1992 Symposium on Some Mathematical Questions in Biology. The Symposium, held in conjunction with the annual meeting of the American Society for Cell Biology (ASCB), was sponsored jointly by the American Mathematical Society (AMS), the Society for Industrial and Applied Mathematics (SIAM), and the Society for Mathematical Biology (SMB), with funding from the National Science Foundation and the ASCB. We would like to express particular thanks to James Maxwell, Donna Salter, and Donna Harmon of the AMS for their help in organizing the conference and publishing the lectures; to Paul Wassarman and Robin Roth of the ASCB for hosting the Symposium; and to Alan Perelson of the SMB for support of a Symposium dinner at which additional talks were presented. We are grateful to the following speakers for excellent short presentations at the dinner and during the sessions: Walter Carrington (University of Massachusetts), Carolyn R. Cho (University of Toronto), Paul DeMilla (Carnegie Mellon University), George Karreman (University of Pennsylvania), Jonathan Sherratt (University of Oxford), James Sneyd (UCLA), and Robert T. Tranquillo (University of Minnesota). Finally, we owe a special debt of gratitude to Patricia Reitemeier for preparing much of this volume.

Byron Goldstein
Carla Wofsy

September 1993

Lectures on Mathematics in the Life Sciences
Volume **24**, 1994

Mathematical Questions in Axonal Transport

Michael C. Reed and Jacob J. Blum

ABSTRACT. The transport of materials in nerve axons gives rise to mathematical models involving linear and nonlinear reaction-hyperbolic partial differential equations. The biological background and significance are discussed, as well as mathematical questions which arise naturally in the analysis of the models.

I. Introduction.

In all cells, proteins, membrane-bound organelles, and other structures (e.g. chromosomes, intermediate filaments, microtubules) are transported from place to place. These transport processes are often highly specific, organized, and proceed at speeds which are much higher than diffusion. Though these transport processes are fundamental to cell maintenance and function, many of the underlying mechanisms, organizational principles, and regulatory features remain unknown.

One set of transport processes which have been studied extensively are those in nerve axons. There are several reasons for this. First of all, protein synthesis occurs in the cell body and then the proteins needed for axonal maintenance and function are shipped down the axons which are typically extremely long and narrow. Thus the transport is essentially one dimensional. This makes both experimental probes and mathematical models easier to construct

---

AMS MOS Subject Classifications 35L45, 92A09. Supported by Grant No. DMS8822449 from the National Science Foundation and Grant No. R01N525191-01A2 from the National Institutes of Health.

and utilize. Secondly, since the transport occurs over such great distances (up to one meter in human sciatic nerve), it is intrinsically interesting to study how it can be maintained and regulated at the molecular level over such great distances. Finally, there is strong evidence that transport is impaired in the presence of a variety of neurotoxins and in many diseases ([10],[11],[17],[18],[19],[20], [33]). The resulting nerve degeneration and dysfuction can be one of the most serious consequences of the disease state.

Since 1984 we have been constructing mathematical models in order to investigate axonal transport by testing hypotheses, matching data, and making predictions. These models are all of the form of initial- boundary value problems for reaction-hyperbolic systems. That is,

$$Lu = f(u). \tag{1}$$

where $u(x,t) =< u_1(x,t), u_2(x,t), ... >$ is a vector of chemical concentrations, L is a linear hyperbolic system of first order operators in one space dimension, and $f(u)$ is a (typically) nonlinear mapping representing the interactions of the various chemical constituents. We will first describe some specific models for fast and slow axonal transport and then turn our attention to mathematical questions that arise naturally in the theory.

II. Fast Transport

In one of the standard experiments used to investigate axonal transport, radiolabelled amino acids are injected into a nerve ganglion. The amino acids are used in protein synthesis in the cell bodies in the ganglion and after several hours a wave of radioactivity consisting of radiolabelled proteins (either individual or packaged as membrane-bound vesicles) begins to propagate down the axon. Figure 1 shows this wave of radioactivity at various times after the injection in goat sciatic nerve. Figure 2 shows the results of similar experiments in garfish olfactory nerve except that several hours after the initial injection the radiolabel not already contained in cell bodies is washed out. Thus, in the experiment in Figure 2, radiolabelled amino acids are available for protein synthesis only for a fixed interval of time and the resulting wave has the (rough) form

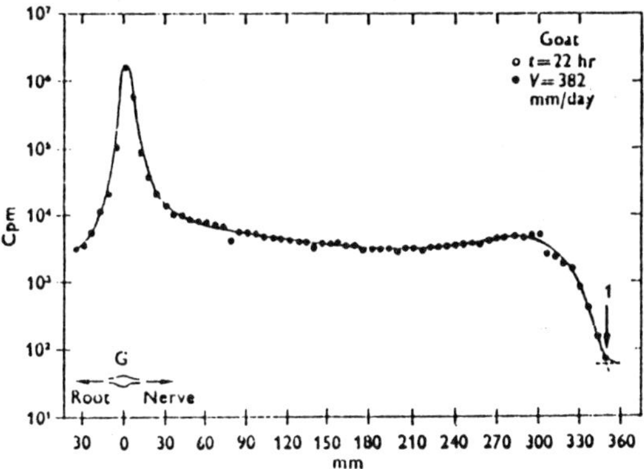

Figure 1: The distribution of radioactive leucine in goat sciatic nerve 22 hours after injection into the ganglion where the cell bodies are located. After approximately one day (22 hours), the axon was cut into small pieces and the radioactivity in each segment was measured as shown on the ordinate. The peak at $x = 0$ represents radioactivity still remaining in the ganglion. The wave front moves at essentially constant velocity and is still quite sharp after 22 hours. The flat plateau behind the advancing wave front shows that radioactivity enters the axon at a constant rate. Taken from [27].

of a pulse travelling down the axon. In both cases, the wave travels at approximately constant velocity down the axon at speeds in the range 0.2 to 1 meter per day. The speeds and exact shapes of the profiles depend on the nerves, animals, and proteins being studied. This is so called fast anterograde transport. Notice the sharp approximately parallel wave fronts which suggest travelling waves. For the large vesicles transported by this system, these speeds are many orders of magnitude faster than diffusion.

Fast transport also occurs in the opposite direction, i.e. towards the soma. Some viruses and growth factors enter the axon by endocytosis at its distal end and are transported towards the soma, i.e. in the retrograde direction. Similarly, proteins that reach the distal end of the axon by slow transport (see below) may be partially degraded and repackaged into multivesicular organelles that are carried by the retrograde transport system. The velocity of transport in the retrograde direction is comparable but somewhat larger than the velocity in the anterograde direction. The mechanochemical

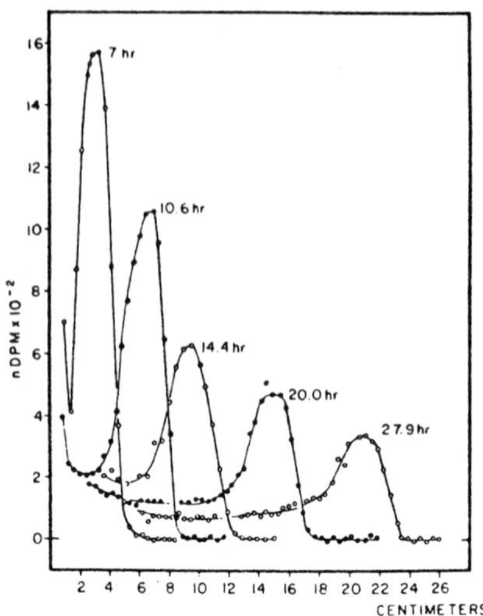

Figure 2: Figure 2 shows the distribution of radioactively labelled proteins in garfish olfactory nerve at various times after injection. In this experiment the cell bodies were severed from the axons at t = 6 hours after which time no new radioactivity entered the axon. Taken from [21].

motor proteins (engines) responsible for the fast anterograde and fast retrograde translocation are termed kinesin and dynein, respectively (see, for example, [34]). In 1985, we hypothesized [6] that in anterograde transport the engines interact reversibly with the vesicular membrane and that these engine-organelle complexes can then bind reversibly to sites on microtubules, long chain polymers oriented parallel to the axon. Almost simultaneously, these hypotheses, indicated schematically in Figure 3, were shown to be correct in intact cells by the technique of video-enhanced light microscopy [1] as well as in experiments in which isolated microtubules were studied (see references in [35]).

It is known that microtubules are asymmetric with a plus and minus end. Kinesin directs the organelle toward the plus end of the microtubule and dynein towards the minus end. Simultaneous attachment of both kinesin and dynein can lead to spontaneous reversals which can be seen in intact cells. For simplicity, we will describe our 1985 model for the anterograde transport system. The model can easily be extended to incorporate both systems simulta-

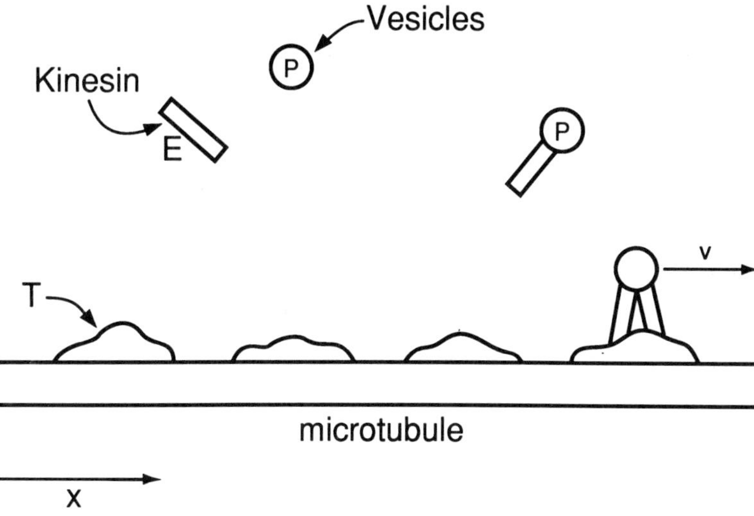

Figure 3: Schematic model of fast transport.

neously. It is useful and interesting to read the article by Garrett Odell in this series, [28], which gave a thoughtful discussion of possible mechanisms for fast transport before these discoveries were made.

Let x denote distance down the axon from the axon hillock which is located at $x = 0$. Denote by $P(x,t), E(x,t), T(x,t), C(x,t)$, and $Q(x,t)$ the concentrations at x at time t of free vesicles, free kinesins, free track sites, vesicle-kinesin compounds, and assembled complexes respectively. The chemical relations among these components,

$$P + nE \underset{k_2}{\overset{k_1}{\rightleftharpoons}} C$$

$$C + T \underset{k_4}{\overset{k_3}{\rightleftharpoons}} Q$$

lead, via the law of mass action, to the system of partial differential equations

$$\frac{\partial P}{\partial t} = -k_1 E^n P + k_2 C \tag{2}$$

$$\frac{\partial C}{\partial t} = k_1 E^n P - k_2 C - k_3 CT + k_4 Q \tag{3}$$

$$\frac{\partial Q}{\partial t} + v\frac{\partial Q}{\partial x} = k_3 CT - k_4 Q \tag{4}$$

for P, C, and Q, where the vD_x term represents the transport of Q at constant velocity v. It is reasonable (in this simplified model) to assume that the total concentrations of kinesins and track sites, both free and bound, are constants E_0 and T_0. Thus, we have the conservation conditions:

$$E_0 = E + nC + nQ \tag{5}$$

$$T_0 = T + Q. \tag{6}$$

One must also specify the concentration of Q at $x = 0$,

$$Q(0,t) = Q_0 \tag{7}$$

which is equivalent to specifiying the amount of transported material, vQ_0A_0, entering the axon per unit time where A_0 is the cross-sectional area. If we solve (5) and (6) for E and T, substitute in (2)-(4), and set the initial conditions,

$$P(x,0) = C(x,0) = Q(x,0) = 0 \tag{8}$$

then (2)-(4), (7), (8) is a well-posed initial-boundary value problem for a semilinear system of hyperbolic partial differential equations. This model represents the propagation of material into an empty axon or a portion of an axon from which vesicles have been removed. Experimentally, this happens in cold block experiments where the temperature is lowered in a short segment of the axon. Transport in that segment stops but continues downstream, effectively emptying the distal portion of the axon.

If we consider a normal axon (no cold block), then to a good first approximation we could consider it to be homogeneous in x and at steady state. This means that at each x the total concentrations of P, E, T, C, and Q are at equilibrium. What we are interested in are the concentrations of radiolabelled P, C, Q, as they propagate into the axon slowly replacing the unlabelled P, C, and Q. In this case one has the same equations (2) -(4) for the concentrations of the radiolabelled P, C, and Q except that now E and T are constants since the kinesins and tracks do not distinguish between radiolabelled and nonradiolabelled P, C, and Q. Thus, in this case (2)-(4) is a linear hyperbolic system with constant coefficients.

The caricature of fast transport described above is much too simple in one respect. It is known that transported material can be irreversibly removed from the transport system, deposited locally, and eventually broken down and returned to the cell body via another (retrograde) fast transport system which moves material in the opposite direction. We can model this by letting $M(x,t)$ be the concentration of deposited material and $D(x,t)$ be the concentration of material in the retrograde transport system. For simplicity, we assume that the kinetics are first order:

$$P(x,t) \overset{k_5}{\rightarrow} M(x,t) \tag{9}$$

$$M(x,t) \overset{k_6}{\rightarrow} D(x,t) \tag{10}$$

In the new system, equation (2) will have an additional term $-k_5P$ and there are 2 new equations,

$$\frac{\partial M}{\partial t} = -k_5 - k_6 M \tag{11}$$

$$\frac{\partial D}{\partial t} - v_1\frac{\partial D}{\partial x} = k_6 M \tag{12}$$

Because of the deposition of material, the axon can not be completely homogeneous but it can be locally at equilibrium. Thus, in this less simple "normal" axon, E and T will not depend on t but they will depend on x and the system (2)-(4), (11)-(12), becomes linear hyperbolic with nonconstant coefficients. These models, linear with constant coefficients, linear with nonconstant coefficients, and semilinear are used in [5], [6], and [30] to investigate various aspects of fast transport. The purpose here was to show how such systems arise naturally in this biological situation.

We want to explain why we were driven by theoretical considerations to our model of the fast transport system even before many of the experimental facts were known. A striking feature of axons is their near homogeneity. Any proposed transport mechanism must explain how this near homogeneity is preserved at great distances from the cell body where protein synthesis takes place. We considered many possible mechanisms but only the ones with translocation and reversible chemical binding could automatically preserve this homogeneity. If a region of the axon is "poor" in some particular protein or vesicle, then, since both association and dissociation rates are driven by concentration, there will be a net movement off the transport mechanism at the protein poor region until the deficit is made up and local equilibrium is achieved. In addition, the computational experiments that we carried out showed that the radioactivity profiles in the model had all of the the most striking features of the experimental data. Finally, the changes that we observed when parameters were varied were consistent with observed differences in profiles in different nerves.

III. Slow Transport and Neuropathies

Long after the the pulse in Figure 2 has passed down the axon, a further pulse emerges from the cell body (see Figure 4) and travels down the axon at a speed roughly one millimeter per day. This is so called slow transport. As in fast transport, the radioactivity profiles keep their shape approximately and travel down the axon at constant velocity, except that now the time scale is days and the space scale is millimeters.

The slow transport system moves neurofilaments, actin microfilaments, tubulin, and several other proteins at speeds of approximately 0.5 to 2 millimeters per day. In most nerves, slow transport has two components. A faster component, SCb, contains actin, myosin-like enzymes, glycolytic enzymes, and a variety of other proteins such as calmodulin and tubulin. The slower component, SCa, contains neurofilaments, tubulin, and some other proteins. Table 1 gives more information on the protein composition and speeds of materials moved by the fast and slow transport systems. It is not known how the slow transport system is organized, what the underlying force-generating mechanism is, or even whether it is one system or two.

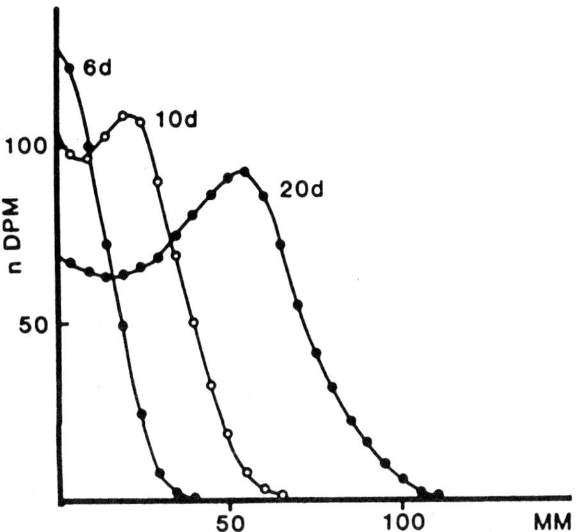

Figure 4: Figure 4 shows slow transport radioactivity profiles in garfish olfactory nerve. Time is now measured in days and distance in millimeters, but the qualitative behavior is the same: relatively sharp wavefronts move at constant velocity and spread slowly. Taken from [13].
Reprinted by permission of Elsevier Science Publishers.

Transport group	Type of structure	Rate	Direction	Protein composition
		Membranous elements		
FC or I	Vesicles, granules (SER?)	250–400 mm/day	Orthograde	Glycoproteins, glycolipids, peptides, catecholamines, acetylcholinesterase
Retrograde	Multivesicular bodies, lamellar bodies	200 mm/day	Retrograde	Lysosomal hydrolases
II	Mitochondria	40 mm/day	Orthograde	Mitochondrial proteins, fodrin
		Cytoplastic elements		
III	Myosin containing complex?	15 mm/day	Orthograde	Myosin-like protein
SCb or IV	Granulo-filamentous matrix	2–5 mm/day	Orthograde	Actin, clathrin, enolase, calmodulin, actin binding protein, fodrin
SCa or V	Microtubule-neuro-filament network	0.2–1 mm/day	Orthograde	Tubulin, neurofilament triplet, tau proteins, fodrin

Table 1: The Table shows the variety of elements moving at different speeds in the fast and slow transport systems. Taken from [22].
Reprinted by permission of Springer-Verlag.

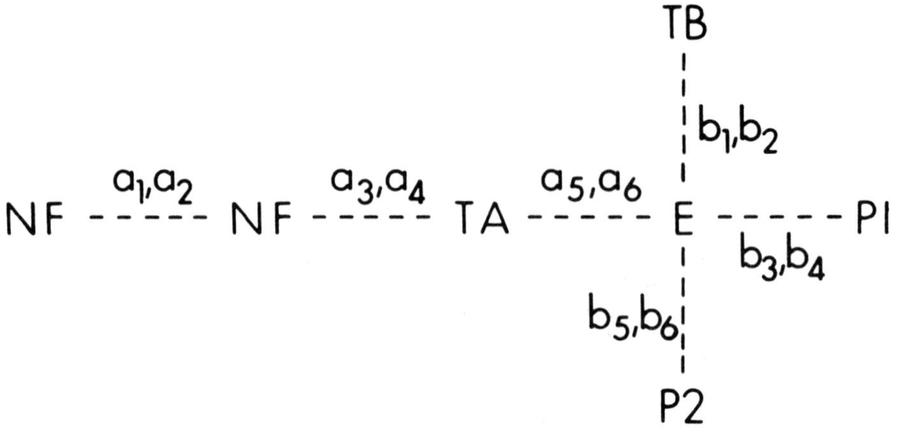

Figure 5: The allowed interactions between components of SCa and SCb in the model are shown by dashed lines. The rate constants shown are the forward and reverse constants for each interaction. Taken from [8].
Reprinted by permission of Wiley-Liss, a division of John Wiley & Sons, Inc.

In 1989 we developed a translocation and reversible binding model for slow transport [5]. This simple model had the following elements: neurofilaments (NF), a motile element (engine, E), two kinds of tubulin (TA and TB), and two cytosolic protein (P_1 and P_2). The possible chemical interactions between these elements are represented by dashed lines (with rate constants) in Figure 5 . Thus, neurofilaments can bind to each other and to TA but not directly to E while the other elements can bind directly to E. When an element is bound to an E, either directly or indirectly, we assume that it moves at constant velocity v. When it is not so bound, it doesn't move. These hypotheses imply that the chemical concentrations (as functions of x and t) corresponding to the various possible bound combinations of these elements satisfy a set of nonlinear partial differential equations such as were described in the fast transport discussion. This system is large and complicated because of the large number of chemical species that one must keep track of, but it can be investigated by machine computation. Figure 6 shows the results of such a machine computation.

It is a common fallacy that there is no point to mathematical

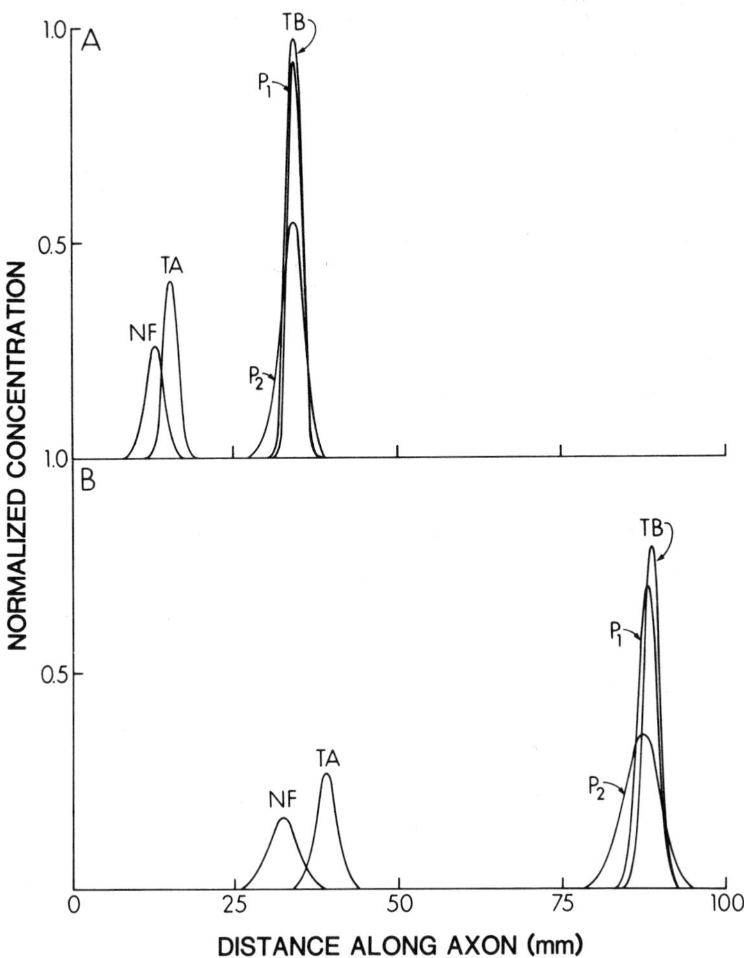

Figure 6: Coherent transport of proteins in SCa and SCb in the model. In each of these simulations the axonal length was 100mm, pulse labeling at the soma was of 3 days duration, the speed, V_0, of translocation when attached to E was 1mm/day. Panels A and B show the results obtained at 40 and 100 days, respectively. Taken from [8].

modelling if the biology is not completely understood. On the contrary, that is when mathematics is particularly useful for working out the consequences of various biological possibilities and testing the results against what is known experimentally. Here are some of the things we learned from this study.

1. Coherent wave-like profiles are observed experimentally for both SCa and SCb. We showed that a single transport mechanism utilizing reversible binding and translocation (and a single velocity v) can produce two groups of coherent profiles travelling at markedly different velocities.

2. In most experiments, SCa travels quite a bit slower than SCb. In the model this occurs naturally because of the relatively weak binding of TA to the motile element suggesting that simple differences in association and dissociation constants could be the cause of the speed differences. Similarly, within SCa the neurofilament pulse typically trails the tubulin pulse in experimental data. In the model this happens naturally because NF binds to the transport mechanism only through TA.

3. Many cytosolic proteins are transported as a coherent group in SCb. We showed using the model that this can remain true whether these proteins bind directly to the transport mechanism or piggyback on each other.

4. The speeds of SCa and SCb vary in different nerves. We showed that in the model speeds depend on total E, NF, and TA concentrations in a manner completely consistent with experimental data in differing nerves.

5. Finally, we were able to use the model to explain an apparent paradox in neurofilament transport rates in the presence of various neurotoxicants such as 2,5-hexanedione (2,5-HD). These neurotoxicants at high doses or prolonged exposure cause the slowing down of neurofilament transport and the eventual formation of neurofilamentous tangles which disrupt the slow transport system. It is thought that the tangles may occur because the neurotoxicant causes an increase in

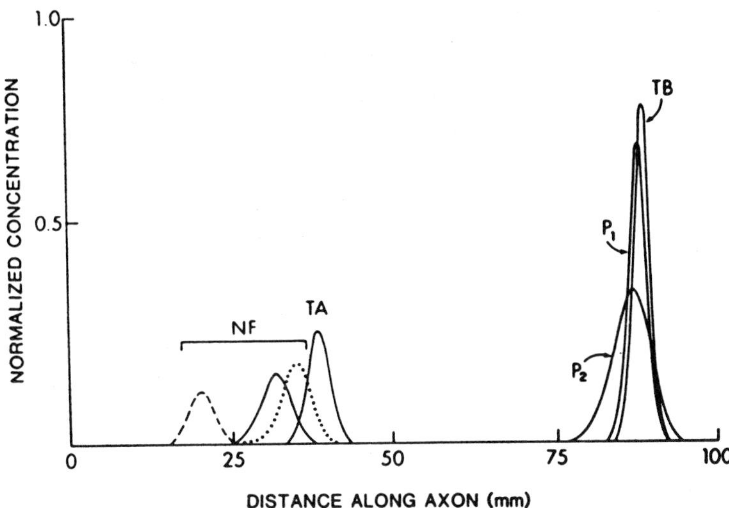

Figure 7: The effect of neurotoxicants that alter $NF-NF$ or $NF-TA$ interactions on the speed of neurofilament transport. The solid line shows the same simulation as in Figure 6. The dotted line shows the speedup of the NF pulse caused by a neurotoxicant which increases $NF-NF$ binding(such as 2,5-HD). The dashed line shows the slowing of the NF pulse caused by a neurotoxicant which interferes with TA - E binding (such as β, β' – *iminodiproprionitrile*). Taken from [8].
Reprinted by permission of Wiley-Liss, a division of John Wiley & Sons, Inc.

neurofilament-neurofilament binding.. It was, therefore, very surprising when various groups reported that at low doses and short times the presence of these neurotoxicants causes the neurofilaments to speed up ([25], [29]). This speedup happens naturally in the model as is shown in the simulations depicted in Figure 7. The intuitive reason is that when the neurofilament-neurofilament binding rate is increased there will be more NF-NF dimers and fewer NF monomers. When one of the NFs in a dimer associates to an E then both are transported, so, on the average the NFs will be transported faster.

It is unlikely that the model is "right". It is a caricature of slow transport which is too simple to explain many important aspects of slow transport and some of our underlying assumptions about the associations of the elements may be wrong. Nevertheless, it is fair to say that the model calculations shed light on the biological situation and, because the model presents a unified conceptual framework, it can serve as basis for discussion and for the design of experiments.

IV. Approximate Travelling Waves.

One of the more interesting mathematical questions that arises in studying the partial differential systems of axonal transport is the question of travelling waves. Much of the experimental data shows travelling waves and the numerical simulations of the models show travelling waves for appropriate choices of the parameters. Yet, in many cases one can show that the system of partial differential equations does not have travelling wave solutions. What is going on here? The equations must have approximate travelling waves in some sense, solutions that look more and more like travelling waves in certain parameter regions, in certain aymptotic limits. To illustrate the question, let us consider this question for the simplest example of the type of system we have been considering. For convenience in referring to [31] we will use lower case letters. Suppose that $p(x,t)$ and $q(x,t)$ are the concentrations of vesicles on and off the transport mechanism, respectively, and that the chemistry is:

$$p \rightleftharpoons q$$

Then the corresponding partial differential equations are:

$$\frac{\partial p}{\partial t} = -k_1 p + k_2 q \tag{13}$$

$$\frac{\partial q}{\partial t} + v\frac{\partial q}{\partial x} = k_1 p - k_2 q \tag{14}$$

For simplicity we will consider the pure initial value problem with the initial conditions:

$$q = \begin{cases} q_0 & x \leq 0 \\ 0 & x > 0 \end{cases} \tag{15}$$

where the pair $< p_0, q_0 >$ is an equilibrium point of the chemical reaction above. That is, at time $t = 0$ all points to the left of $x = 0$

have the same amount of radioactivity at equilibrium between p and q. There is no radioactivity to the right. If there were a travelling wave solution $p = f(x-at), q = g(x-at)$, then f and g would satisfy

$$-af'(t) = -k_1 f(t) + k_2 g(t) \tag{16}$$
$$(v-a)g'(t) = k_1 f(t) - k_2 g(t) \tag{17}$$

and we would have $f(t) \to p_0$ and $g(t) \to q_0$ as $t \to -\infty$, and $f(t) \to 0$, $g(t) \to 0$ as $t \to \infty$. Since the system is linear, the solutions are linear combinations of exponentials and these conditions cannot hold. Thus there are no travelling wave solutions. Yet numerical calculations show approximate travelling waves. Where are they? Intuitively they should occur for large k_1 and large k_2. The left hand side of (13)-(14) gives the transport and the right hand side the local adjustment of the chemistry. As k_1 and k_2 get larger the chemistry adjusts more quickly towards equilibrium at each local point and thus the wave should keep its shape better. To try out this idea we rewrite equations (13)-(14) with an ε on the left hand sides:

$$\varepsilon \partial_t p = -k_1 p + k_2 q \tag{18}$$
$$\varepsilon(\partial_t + v\partial_x) q = k_1 p - k_2 q \tag{19}$$

Equivalently to scaling k_1 and k_2 larger, we hold k_1 and k_2 fixed and ask what happens to the solution as we let $\varepsilon \to 0$. For simplicity of notation we are writing ∂_t instead of $\frac{\partial}{\partial t}$ and ∂_x instead of $\frac{\partial}{\partial x}$. Differentiating (18) with respect to t and using (19) we find:

$$\begin{aligned} \varepsilon(\partial_t + v\partial_x)\partial_t q &= k_1 \partial_t p - k_2 \partial_t q \\ &= k_1 \varepsilon^{-1}(-k_1 p + k_2 q) - k_2 \partial_t q \\ &= -k_1(\partial_t + v\partial_x) q - k_2 \partial_t q \end{aligned}$$

or,

$$\varepsilon(\partial_t + v\partial_x)\partial_t q + k_1(\partial_t + v\partial_x) q + k_2 \partial_t q = 0 \tag{20}$$

To study the behavior of this equation for ε small, we make a change of dependent and independent variables,

$$t \longrightarrow t\,, \qquad x \longrightarrow \quad y \equiv \frac{x-at}{\varepsilon^{1/2}}\,, \qquad q \longrightarrow Q$$

That is,

$$q(x,t) = Q(y,t) = Q(\frac{x-at}{\varepsilon^{1/2}}, t)$$

and a short calculation shows that (20) implies that Q satisfies:

$$\begin{aligned}&\varepsilon(\partial_t + \varepsilon^{-1/2}(v-a)\partial_y)(\partial_t - \varepsilon^{-1/2}u\partial_y)Q \\ &\quad + k_1(\partial_t + \varepsilon^{-1/2}(v-a)\partial_y)Q + k_2(\partial_t - \varepsilon^{-1/2}u\partial_y)Q = 0\end{aligned}$$

We first require that the terms of order $\varepsilon^{-1/2}$ vanish which yields

$$(k_1(v-a) - k_2)\partial_y Q = 0$$

or

$$a = \frac{k_1 v}{(k_1 + k_2)} \tag{21}$$

If the axon is locally at equilibrium then the chemistry implies that a vesicle will spend on the average $k_1/(k_1+k_2)$ part of the time in the form p where it is attached to the transport mechanism and travels at speed v. Thus, it makes sense that the speed of the radioactive wave is given by this expression for a.

If we now require that the terms of order ε^0 vanish, we get:

$$(k_1 + k_2)\partial_t Q - (v-a)a\partial_y^2 Q = 0$$

or, using (21),

$$\partial_t Q - \kappa \partial_y^2 Q = 0 \tag{22}$$

(23)

where

$$\kappa = \frac{k_1 k_2}{(k_1 + k_2)^3} \tag{24}$$

Thus, when ε is small, q(x,t) is approximately given by:

$$q(x,t) \sim Q(\frac{x-at}{\varepsilon^{1/2}},t)$$

where Q satisfies the *heat* equation, the initial condition (15), and a and κ are given by (21) and (24). The formula for p(x,t) is similar. There is the approximate travelling wave! a is the speed, ε governs the sharpness of the front, and κ governs how the front spreads as it travels. This simple analysis is carried out in detail for the general *linear* reaction-hyperbolic equation in one space dimension in [31]. The nonlinear cases remain open. In the linear case , these results can also be obtained by a Markov chain analysis [23] and Chen et al. [15] have recently shown in a general context why one should expect a dissipative equation near the asymptotic limit even though the underlying partial differential equation in hyperbolic.

Notice that in the linear two by two case discussed above, (21) and (24) can easily be solved to express k_1 and k_2 by simple formulas in terms of a and κ. a is the speed of the travelling wave and κ is the (diffusion controlled) spreading and both of these macroscopic quantities can easily be estimated from the experimental radioactivity curves. The formulas then allow one to *compute* k_1 and k_2, the (effective) association and dissociation constants which are microscopic quantities not easily measured. This illustrates one important way in which mathematics can help biology, by relating, through theory and calculation, macroscopic and measurable quantities to important microscopic quantities. Of course, if the chemistry is more complicated then it is not entirely determined by a and κ and more difficult analyses of the data and the relationship to the chemistry need to be carried out.

V. Existence and A Priori Estimates.

There are many difficult and interesting questions in the existence and uniqueness theory for systems of the form

$$Lu = f(u)$$

where L is a linear hyperbolic partial differential operator and f is a nonlinear function reflecting the chemical interactions of the constituents. If the chemistry is nice (e.g. a unique, asymptotically stable, global equilibrium for each fixed total mass) then one would expect that the existence and uniqueness theory should be easy. The left sides of the equations transport and the right sides adjust the chemistry locally. It seems that one should be able to use a Trotter product (i.e. splitting) method to get existence, uniqueness and estimates. Such methods are good for intuition (and numerical calculation) but so far have not been successfully used for existence theory. A well known example is the Broadwell Model which is a three velocity caricature of the Boltzmann equation. The "chemistry" is well behaved and the linear part couldn't be simpler. Yet existence, uniqueness, and proofs of asymptotics were very difficult and used many special properties of the right hand sides (see for example [2] or [3]).

A excellent start on these existence and uniqueness problems in one space dimension has been made by Danielle Carr who completed her thesis at Duke in 1992, [14]. She completely analyzes the two characteristic nonlinear case, has some general results for the n characteristic case, and proves existence, uniqueness, and estimates for particular examples important in axonal transport by using special properties of the right hand sides.

References

1. Allen, R. D., Weiss, D. G., Hayden, J. H., Brown, D. T., Fujiwake, H., and Simpson, M. 1985. Gliding movement of and bidirectional transport along single native microtubules from squid axoplasm: Evidence for an active role of microtubules in cytoplasmic transport. J. Cell Biol. 100:1736–1752.

2. Arkeryd, L. and Illner, R., 1993. The Broadwell Model in a Box: Strong L^1 convergence to Equilibrium. SIAM J. Appl. Math., in press.

3. Beale, J.T., Large Time Behavior of the Broadwell Model of a Discrete Velocity Gas, 1985. Commun. Math. Phys. 102: pp. 217–235.

4. Bloom, G.S., Richards, B.W., Leopold, P.L., Ritchey, D.M., and Brady, S.T., 1993. GTPγS inhibits organelle transport along axonal microtubules, J. Cell Biol. 120: 467–476.

5. Blum, J.J., Carr, D.D., and Reed, M.C., 1992. Theoretical Analysis of Lipid Transport in Sciatic Nerve. Biochim. Biophys. Acta 1125: 313–320.

6. Blum, J.J. and Reed, M.C. 1985. A model for fast axonal transport. Cell Motility 5:507–527.

7. Blum, J.J. and Reed, M.C. 1988. The transport of organelles in axons. Math Biosci. 90:233–245.

8. Blum, J.J. and Reed, M.C. 1989. A model for slow axonal transport and its application to neurofilamentous neuropathies. Cell Motility Cytoskel. 12:53–65.

9. Brady, S.T. and Lasek, R.J. 1982. Axonal transport: A cell biological method for studying proteins that associate with the cytoskeleton. Methods in Cell Biol. 25:365–398.

10. Braendegaard, H. and Sidenius, P. 1986. Anterograde components of axonal transport in motor and sensory nerves in experimental 2,5- hexanedione neuropathy. J. Neurochem. 47:31–37.

11. Breuer, A. C., Lynn, M. P., Atkinson, M. B., Chou, S. M., Wilbourn, A. J., Marks, K. E., Culver, J. E. and Fleegler, E. J. 1987. Fast axonal transport in amyotrophic lateral sclerosis: an intra axonal organelle traffic analysis. Neurology 37:738–748.

12. Burgoyne, R. D. (editor) 1991. The Neuronal Cytoskeleton. Wiley-Liss, Inc., New York, 334pp.

13. Cancalon, P., 1983. Influence of Temperature on Slow Flow in Populations of Regenerating Axons with Different Elongation Velocities. Devel. Brain Res. 9: pp. 279–289.

14. Carr, D., 1992, Reaction-Hyperbolic Systems in One Space Dimension. Duke University Thesis, 1992.

15. Chen, G.-Q., Levermore, C.D., and Liu, T.-P., 1992. Hyperbolic Conservation Laws with Stiff Relaxation Terms and Entropy, MSRI preprint.

16. Garner, J.A. and Lasek, R.J. 1982. Cohesive axonal transport of the slow component complex of polypeptides. J. Neurosci. 2:1824–1835.

17. Gold, B. G. 1987. The pathophysiology of proximal neurofilamentous giant axonal swellings: implications for the pathogenesis of amyotrophic lateral sclerosis. Toxicology 46:125–139.

18. Graham, D. G., Szakal Quin, G., Priest, J. W. and Anthony, D. C. 1984. In vitro evidence that covalent cross-linking of neurofilaments occurs in diketone neuropathy. Proc. Natl. Acad. Sci. USA 81:4979–4982.

19. Griffin, J. W., Anthony, D. C., Fahnestock, K. E., Hoffman, P. N. and Graham, D. G. 1984. 3,4- Dimethyl 2,5- hexanedione impairs the axonal transport of neurofilament proteins. J. Neurosci. 4:1516–1526.

20. Griffin, J.W., and Watson, D.F., 1988. Axonal Transport in Neurological Disease. Ann. Neurol. 23: pp. 3–13.

21. Gross, G.W. and Beidler, L.M., 1975. A Quantitative Analysis of Isotope Concentration Profiles and Rapid Transport Velocities in the C-fibers of the Garfish Olfactory Nerve. J. Neurobiol. 6: pp. 213–232.

22. Lasek, R.J. and Brady, S.T., 1982. The Structural Hypothesis of Axonal transport: Two Classes of Moving Elements, in *Axoplasmic transport* (ed. D. Weiss), Springer-Verlag, New York.

23. Lawler, G., private communication.

24. McLane, J. A. 1987. Decreased axonal transport in rat nerve following acute and chronic ethanol exposure. Alcohol 4:385–389.

25. Monoco, S., Autilio-Gambetti, L., Zabel, D., and Gambetti, P. 1985. Giant axonal neuropathy: acceleration of neurofilament transport in optic Axons. Proc. Natl. Acad. Sci. USA 82: pp. 920–924.

26. Medori, R., Autilio- Gambetti, L., Monaco, S. and Gambetti, P. 1985. Experimental diabetic neuropathy: impairment of slow transport with changes in axonal cross sectional area. Prod. Natl. Acad. Sci. USA 82:7716–7720.

27. Ochs, S., 1972. Rate of Fast Axoplasmic Transport in Mammalian Nerve Fibers. J. Physiol. 227: pp. 627–645.

28. Odell, G. M., 1977, Theories of Axonal Transport. In: *Lectures on Mathematics in the Life Sciences*, American Math. Soc., Providence, pp.141–186.

29. Pyle, S. J. , Amarnath, V., Graham, D.G. and Anthony, D.C. 1991. Acceleration of neurofilament transport following 2,5-Hexanedione intoxication is a persistent response of the axon. J. Neuropathol. Exp. Neurol. 50:321.

30. Reed, M.C. and Blum, J.J. 1986. Theoretical analysis of radioactivity profiles during fast axonal transport: Effects of deposition and turnover. Cell Motility Cytoskel. 6:620–627.

31. Reed, M.C., Venakides, S., and Blum, J.J. 1990. Approximate travelling waves in linear reaction-hyperbolic equations. SIAM J. Appl. Math. 50:167–180.

32. Sayre, L.M., Gambetti, A., and Gambetti, P., 1985. Pathogenesis of experimental giant neurofilamentous axonopathies: a unified hypothesis based on chemical modification of neurofilaments. Brain Res. Rev. 10: pp. 69—83.

33. Schlaepfer, W.W., 1987. Neurofilaments: structure, metabolism, and implications in disease. J. Neuropathol. Exp. Neurology 46: pp. 117-129.

34. Sheetz, M.P. and Martenson, C.N. 1991. Axonal transport: Beyond kinesin and cytoplasmic dynein. Curr. Opinions Neurobiol. 1:393–398.

35. Vale, R.D., Malik, F. and Brown, D., 1992. Directional instability of microtubule transport in the presence of kinesin and dynein, two opposite polarity motor proteins, J. Cell Biol. 119: pp. 1589–1596.

Department of Mathematics, Duke University, Box 90320, Durham, NC 27708-0320
E-mail address: reed@math.duke.edu

Department of Cell Biology, Duke University Medical Center, Durham, NC 27708

Lectures on Mathematics in the Life Sciences
Volume **24**, 1994

CURVATURE SEGREGATION OF PROTEINS IN THE GOLGI

ERIC ODELL, GEORGE OSTER

ABSTRACT. Membrane proteins in the Golgi cisternae are sequestered at the peripheral regions where they are packaged into secretory vesicles. We propose a mechanism for this process based on the mechanical coupling of membrane curvature to diffusion of transmembrane proteins.

1. INTRODUCTION

Biological membranes are two-dimensional fluids containing proteins which may constitute nearly 50% of the surface area. This mixture is far from homogeneous; lipids and proteins cluster in islands that relate to their cellular functions. There are several proposed mechanisms for segregating membrane proteins into distinct domains, including local phase transitions, junctional barriers, and binding to cytoskeletal proteins [18, 26]. We shall explore the possibility that mechanical coupling of membrane proteins to the bilayer curvature influences the spatial distribution of proteins. As we shall demonstrate, proteins of different shapes tend to migrate to—and create—membrane regions of complementary shape. Table 1 gives a sampling of proteins that appear to localize in zones of characteristic curvature.

Golgi stacks consist of flattened discoid cisternae with regions of high curvature at the rim ($\approx 1/50\,\mathrm{nm}$). Vesicles bud from the rims of the cisternae and move to the next cisterna or, when they reach the trans cisterna, to the plasma membrane. This sequential process concentrates

1991 *Mathematics Subject Classification.* Primary 92C10.
EO was supported by American Cancer Society Grant # ACS-CD497 to H.-P. Moore. GFO was supported by NSF Grant FD92-20719.
The detailed version of this paper will be submitted for publication elsewhere.

secretory proteins and segregates them from proteins that remain resident in the Golgi. Here we examine the process by which proteins may localize at the Golgi rim, preparatory to vesicle budding.

Our model builds on the analyses of Markin and of Leibler [15, 13]. Markin studied the equilibrium shapes of membranes within which are embedded proteins of different shapes; his static model did not evaluate the rates of membrane deformation and protein aggregation. Leibler's approach was dynamic; however, he modeled the interaction between membrane and proteins via a phenomenological coupling parameter, and his equations were linear, and so restricted to small membrane deformations. In order to extend these analyses we compute the stress distribution at the interface of the protein and the lipid bilayer. This approach yields an explicit formula for membrane-protein coupling that applies to proteins of arbitrary shape. Moreover, the equations derived in this fashion allow us to model the nonlinear, coupled dynamics of membranes containing proteins.

2. Curvature flux

Superimposed on random Brownian motion, membrane proteins experience a drift velocity:

$$\boldsymbol{v}_d = \frac{D}{k_B T} \boldsymbol{F}_p \tag{1}$$

where $\boldsymbol{F}_p$ is the force on the protein due to stress in the membrane. To derive an expression for $\boldsymbol{F}_p$ we make the following assumptions. First, we characterize the shape of a membrane protein by a cone of mean radius r_p and curvature $\kappa_p = \phi/r_p$, where ϕ is the half-angle of the cone (c.f. Figure 1). By integrating the stress in the bilayer over the surface of the protein we derive the following expression:

$$\boldsymbol{F}_p = 2\pi r_p^2 \mathcal{B}(\kappa_p - \bar{\kappa})\nabla(2\bar{\kappa} - \kappa_s) \tag{2}$$

where $\bar{\kappa}$ is the mean curvature of the membrane at the location of the protein and $\mathcal{B}$ is the bending modlulus of the membrane. Assuming the lipid monolayers are symmetric and contain no intrinsic curvature, the spontaneous curvature, κ_s, is the protein curvature weighted by its local fractional concentration; i.e. $C^* = C/C_{max}$ is the ratio of local protein concentration at position $\boldsymbol{x}$ and time t to pure protein concentration C_{max}:

$$\kappa_s = \kappa_p C^* = \kappa_p \frac{C}{C_{max}} \tag{3}$$

Thus a protein experiences a force proportional to the mismatch between the protein shape and local membrane curvature and the gradient in membrane curvature.

The protein flux is the sum of random diffusive motion superimposed on the drift velocity:

$$(4) \qquad \boldsymbol{J}(x) = -D\nabla C + \boldsymbol{v}_d C$$

Substituting equations (3) and (2) into the preceding yields the convective flux due to the local composition and shape of the membrane:

$$(5) \qquad \boldsymbol{J}_c = C[D\alpha(\kappa_p - \bar{\kappa})(2\nabla\bar{\kappa} - \frac{\kappa_p}{C_{max}}\nabla C)]$$

where

$$(6) \qquad \alpha \equiv 2\pi r_p^2 \frac{\mathcal{B}}{k_B T}$$

The equation of motion for proteins within the membrane is obtained by inserting the flux expression (4) into the conservation equation, $\partial C/\partial t = -\nabla \cdot \boldsymbol{J}$:

$$(7) \qquad \frac{\partial C}{\partial t} = D\nabla^2 C - \nabla \cdot \boldsymbol{J}_c$$

2.1. Reduction to one dimension: cylindrical surfaces. For a surface of *fixed* shape we can solve for the equilibrium concentration distribution by setting $\boldsymbol{J} = 0$. This yields the differential equation:

$$(8) \qquad [1 + \alpha\kappa_p(\kappa_p - \bar{\kappa})\frac{C}{C_{max}}]\nabla C + 2\alpha(\bar{\kappa} - \kappa_p)C\nabla\bar{\kappa} = 0$$

Figure 2 is a numerical solution of the concentration enrichment on a cylindrical membrane of fixed elliptic cross section whose extremal curvature matches a protein curvature of $\kappa_p = 0.2\,\mathrm{nm}^{-1}$ and has a bending modulus of $\mathcal{B} = 100\,\mathrm{pN\cdot nm}$. Clearly, a considerable equilibrium enrichment of proteins at the rim is possible. An estimate of the effect of the curvature flux on the aggregation rate of proteins can be obtained by computing the mean first passage time for Brownian particles to diffuse to the cisternal edge.[1] This time is decreased by an order of magnitude over that of a flat membrane.

In analyzing the diffusive bias towards the Golgi rim we have assumed

---

[1]The mean first passage time, $\mathcal{T}(0, L)$ for a particle released at the origin to reach the edge at $x = L$ is obtained by solving [31]:

$$D\frac{\partial^2 \mathcal{T}}{\partial x^2} + \frac{D\boldsymbol{F}_p}{k_B T}\frac{\partial \mathcal{T}}{\partial t} = -1 \,, \quad \frac{\partial \mathcal{T}}{\partial x} = 0 \,, \quad \mathcal{T}(L) = 0$$

Averaging the solution over [0,L] yields the mean time for particles to diffuse to the rim.

the cisternal shape is maintained by the elasticity of the lumenal matrix and the structural microtubules necessary for Golgi integrity [23, 28, 30]. The shape of the Golgi cisternae suggests the lumenal matrix may be a dehydrated gel analogous to that found in the lumen of several types of secretory granules that bud from the trans-Golgi network [6, 29]. Several arguments support this. First, osmotically deswelled bilayer vesicles assume cisternal shapes corresponding to a minimum of the bilayer bending energy [24]. Second, the Golgi membrane is rich in phospholipids, cholesterol and glycolipids which are synthesized on the cytoplasmic leaflet. There must be lipid flippases since these lipids are added to secretory lipoproteins in the lumen. The presence of flippases would appear to rule out that the shape of the Golgi is driven by lipid generated bilayer couples since interbilayer stresses and curvature mismatches could be relieved by lipid flipping. Third, monensin, a carboxylic ionophore, equilibrates $Na^+, K^+,$ and H^+ across the Golgi membrane, and induces the cisternae to swell to a spherical shape within seconds [28]. This response to ionic equilibration is reminiscent of the rapid swelling of mucin secretory granules upon fusion with the plasma membrane [29]. In that system, the vesicle contents are maintained in a deswelled condition by sequestered calcium; exchange of calcium for monovalent cations upon fusion leads to rapid swelling and exocytosis. Thus the Golgi lumenal matrix may be composed of a gel, perhaps proteoglycan, which is maintained in a deswelled state by its internal ionic composition.[2] It is worth noting that in the swelled configuration Golgi processes are slowed or halted. One cause for this may be that the spherical shape (i.e. constant curvature) is not conducive to curvature sorting and vesiculation.

3. THE EFFECT OF PROTEIN SHAPE ON MEMBRANE SHAPE

Equations governing membrane shape can be derived in several ways. The usual approach for small deformations is to construct an elastic energy functional comprising the bending and stretching energies. The corresponding Euler-Lagrange equation is a fourth order equation for the normal deformation similar to that found in elastic plate theory [13, 27]. We derive the equations of motion by a direct stress balance; details will be presented elsewhere.

Denote by $u(\boldsymbol{x}, t)$ the normal displacement field, $\boldsymbol{M}(\boldsymbol{x}, t)$ the moment field, and $\boldsymbol{T}(\boldsymbol{x})$ the tension field at point $\boldsymbol{x}$ on the membrane. The normal

[2] For a discussion of ionic gel swelling in the context of cell protrusion, see [21, 22].

component of membrane motion is:

$$\zeta v = \underbrace{\nabla^2 \boldsymbol{M}}_{bending} - \underbrace{(\kappa_1, \kappa_2) \cdot \boldsymbol{T}}_{stretching} - \underbrace{Ku}_{\substack{lumenal \\ elasticity}} \tag{9}$$

where v is the normal velocity of the membrane, κ_1, κ_2 are the principle bilayer curvatures, and K is an elastic modulus characterizing the lumenal matrix. In this approximation we have neglected internal membrane viscosity compared with the frictional drag of the membrane with the aqueous environment, as contained in the drag coefficient, ζ. The tangential equations of motion are the equilibrium relations:

$$\nabla \boldsymbol{T} + \begin{pmatrix} \kappa_1 & 0 \\ 0 & \kappa_2 \end{pmatrix} \nabla \boldsymbol{M} = 0 \tag{10}$$

To close the system we must express the moment field, $\boldsymbol{M}$, in terms of the curvature field, (κ_1, κ_2). To first order this can be expressed as:

$$\boldsymbol{M} = \mathcal{B}(2\bar{\kappa} - \kappa_s) \tag{11}$$

Inserting equations (11) and (3) into (9) yields the membrane equation of motion in terms of the curvature field, (κ_1, κ_2), and protein concentration, C:

$$\zeta v = \mathcal{B} \nabla^2 [2\bar{\kappa} - \kappa_p \frac{C}{C_{max}}] - (\kappa_1, \kappa_2) \cdot \boldsymbol{T} - Ku \tag{12}$$

Equations (12) and (7) describe how a membrane deforms in response to the distribution of proteins, and how proteins move in response to membrane shape. The dynamics are complex and here we shall only explore a one dimensional stability analysis which demonstrates how proteins aggregate in regions of membrane curvature that best accommodate their intrinsic shapes.[3] We shall present numerical simulations of the nonlinear equations elsewhere.

3.1. Budding of secretory vesicles. Secretory granules bud almost exclusively from the Golgi rims, and in the case of the trans-Golgi network, the tips of cisternal tubules. According to our analysis, these regions will contain the highest concentration of membrane proteins whose shape matches the local curvature. It is tempting to speculate that aggregation is a factor triggering, or permitting, vesiculation. We can use the curvature-aggregation equations to estimate the spacing of vesicle budding on the rim of a cisterna. Since the circumferential curvature around

[3]Lee Segel has suggested we call the tendency of proteins to accumulate in regions of specific curvature "campylotaxis" (*campylos* = curvature)

a cisterna is much less than the azimuthal curvature, we shall model the initiation of vesicle budding by examining the distribution of proteins around the rim of a cisterna, thus reducing the model equations to the single dimension of distance along the rim. While there is evidence the bending modulus decreases with increasing curvature [9], for the purposes of a linear stability analysis we assume the bending modulus $\mathcal{B}$ is constant, and curvature $\kappa \approx \frac{\partial^2 u}{\partial x^2}$ where $u(x,t)$ is the normal displacement field. The linearized membrane equations reduce to:

$$\frac{\partial C}{\partial t} = D[(1+\hat{\alpha}\phi)\frac{\partial^2 C}{\partial x^2} - \hat{\alpha}\frac{\partial^4 u}{\partial x^4}] \tag{13}$$

$$\zeta\frac{\partial u}{\partial t} = \mathcal{B}(\frac{\partial^4 u}{\partial x^4} - \phi\frac{\partial^2 C}{\partial x^2}) - T\frac{\partial^2 u}{\partial x^2} - Ku \tag{14}$$

where:

$$\hat{\alpha} \equiv 2\pi r_p^2 \kappa_p C_o \frac{\mathcal{B}}{k_B T} \tag{15}$$

These equations determine the initial stability of the uniform, constant curvature and concentration state; that is, whether a uniform distribution of proteins on a flat membrane will remain stable, or commence to aggregate in patches of proteins which create curved regions. Indeed, the pair of equations exhibit a diffusive instability near the size of a vesicle (diameter $\approx$ 100 nm) [13, 19]. The result of the linear stability analysis is given in Figure (3) where a plot of the dispersion relation yields the mean vesicle spacing expected around the cisterna rim as a function of tension in the membrane.

4. Discussion

In many biological membranes, the surface area occupied by proteins is about the same as that occupied by lipid molecules so that proteins are separated by but a few lipids [3, 16]. Of these intervening lipids at least one layer, and perhaps more, associate closely with proteins; these lipids form semicrystalline regions whose mobility is severely restricted compared to the interstitial lipids. This high packing probably affects the diffusional mobility of membrane proteins [1] as well as the shape of the membrane. We have shown that elastic interactions between lipid bilayers and mobile proteins can lead to aggregations of proteins in regions of membrane curvature that best match the protein geometry.

One consequence of protein aggregation at the rim could be the initiation of vesicle budding. Lipowsky has proposed a simple and appealing mechanism for this [14]. If a region of the bilayer segregates into two domains with different interfacial tensions, line tension between the boundary of the two regions can come to dominate the bending rigidity of the bilayer and drive budding of a spherical vesicle. Lipowsky assumed that abutting lipid domains give rise to this effect, but it is equally plausible that an aggregation of membrane proteins could drive budding [19, 20]. Note also that the neck of a budding vesicle, because of its negative mean curvature, presents an increasing energy barrier to diffusing wedge-shaped proteins, excluding some from the bud and trapping others in it, according to their molecular shape.

It is tempting to generalize our conclusions beyond the Golgi setting. Many membrane associated proteins undergo conformational changes, especially accompanying phosphorylation and dephosphorylation. These reactions may alter the molecular geometry sufficiently to change the protein's local influence on membrane curvature. For example, protein kinase C associates with about a dozen phosphatidylserines, charged and wedge-shaped lipids, on the cytoplasmic leaflet of the plasma membrane. This highly asymmetric aggregate could buckle the membrane, forming a hillock that could act as a curvature trap for substrates or, in the act of forming, aid in attracting the requisite phosphatidylserine moieties. This may be a general phenomenon: membrane proteins surrounding themselves with specific lipids which not only activate enzymatic activity, but create an aggregate whose shape is conducive to curvature sorting. Similarly, curvature trapping may play a role in capturing membrane proteins during coated pit and vesicle formation in the plasma membrane and in sequestering receptors in the tubular arms of endosomes, CURL [8]. From this viewpoint, membrane topography may facilitate membrane reactions.

Protein	***Reference***
Coatamers on the rim of Golgi	[12]
Proton pumps on the flats of contracting vacuoles	[10]
Virus spikes on filopodia	[17]
Actin nucleation sites at the rims of lamellipodia	[5]
Diacylglycerol nucleating actin assembly	[25]
Receptors clustering in coated pits	[2]
GPI anchored proteins in caveolae (e.g. folate receptor)	[11]
Calcium pumps in caveolae	[7]
rab 8 proteins at the tips of lamellae & ruffles of fibroblasts	[4]
Fusagens at the rims of phagocytotic ruffles	J. Heuser (pers. comm.)

TABLE 1. Curvature dependant localization of membrane components

Symbol	*Description*	*Units*
$\mathcal{B}$	Bending modulus of bilayer	pN·nm
C	Fractional protein concentration	nm^{-1}
C_o	Average fractional concentration	nm^{-1}
D	Diffusion constant for proteins in membrane	$\text{nm}^2 \cdot \text{sec}^{-1}$
$\mathcal{E}$	Energy applied to protein	pN·nm
$\boldsymbol{F}_p$	Force applied to protein	pN
$\boldsymbol{J}$	Flux of membrane immersed proteins	nm^{-1}
k_BT	Thermal energy	pN·nm
K	Lumenal elastic modulus	$\text{pN} \cdot \text{nm}^{-3}$
M	Bending moment of bilayer	pN
r_p	Average radius of wedge shaped protein	nm
t	Time	sec
$\mathcal{T}$	First passage time	sec
T	Tension of bilayer	$\text{pN} \cdot \text{nm}^{-1}$
u	Displacement of membrane	nm
v	Normal membrane velocity	$\text{nm} \cdot \text{sec}^{-1}$
$\boldsymbol{x}$	Point on membrane	nm
x	Position along cylindrical membrane	nm
α	$2\pi r_p^2 \mathcal{B}/k_BT$	nm^2
$\hat{\alpha}$	$2\pi r_p^2 \kappa_p C_o \mathcal{B}/k_BT$	dimensionless
ϕ	Half angle of wedge shaped protein	radians
κ_p	Curvature of wedge shaped protein	nm^{-1}
κ_l	Curvature of pure lipid membrane	nm^{-1}
$\bar{\kappa}$	Mean curvature of membrane at $\boldsymbol{x}$	nm^{-1}
κ_s	Spontaneous curvature of monolayer	nm^{-1}
ζ	Membrane drag coefficient	$\text{pN} \cdot \text{sec} \cdot \text{nm}^{-3}$

TABLE 2. List of Symbols

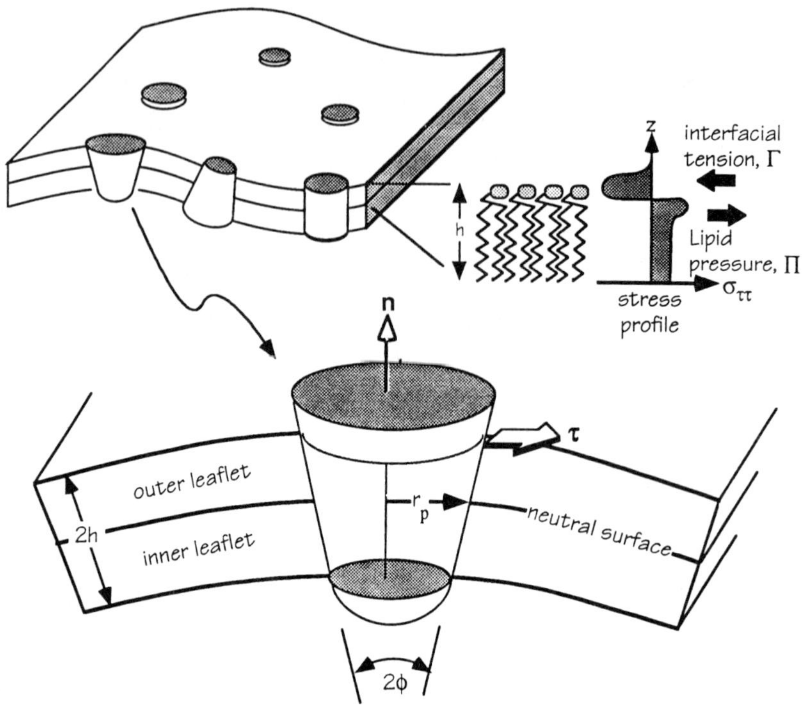

FIGURE 1
Model membrane of two monolayers of thickness h which may slide past one another. The axial stress profile in the plane of the membrane, $\sigma_{\tau\tau}$, is shown schematically to the right. $\sigma_{\tau\tau}$ is the sum of the interfacial tension, Γ, generated by the aqueous environment and the monolayer pressure, Π, due to repulsive forces between lipid molecules. Proteins are modeled as conical wedges with apical angle 2ϕ and radius r_p at the neutral surface. $\{\boldsymbol{\tau}, \boldsymbol{n}\}$ denote unit tangent and normal to the membrane.

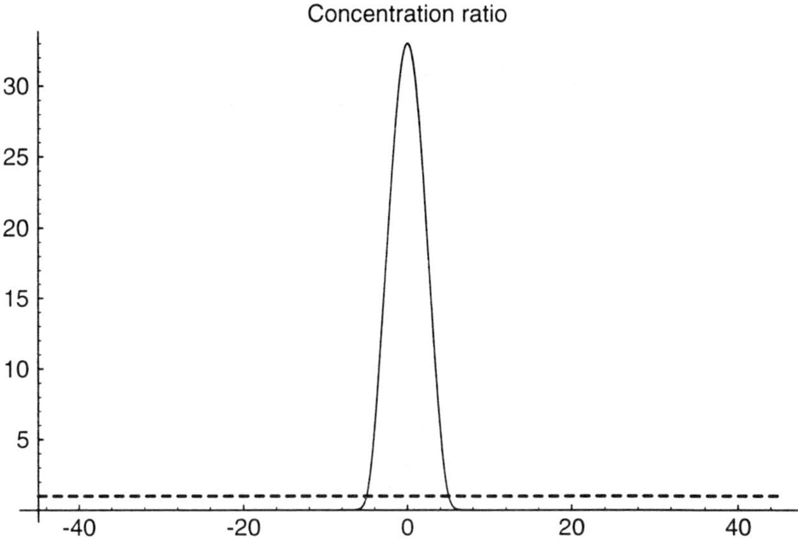

FIGURE 2
The concentration enrichment on an elliptical cylinder due to membrane curvature. The ellipse has principle diameters 1000 nm and 100 nm. Distance along the ellipse is measured by the polar angle from the rim, and the concentration ratio is measured relative to the initial uniform distribution of proteins (the dashed line represents the initial, uniformly distributed concentration ratio of 1).

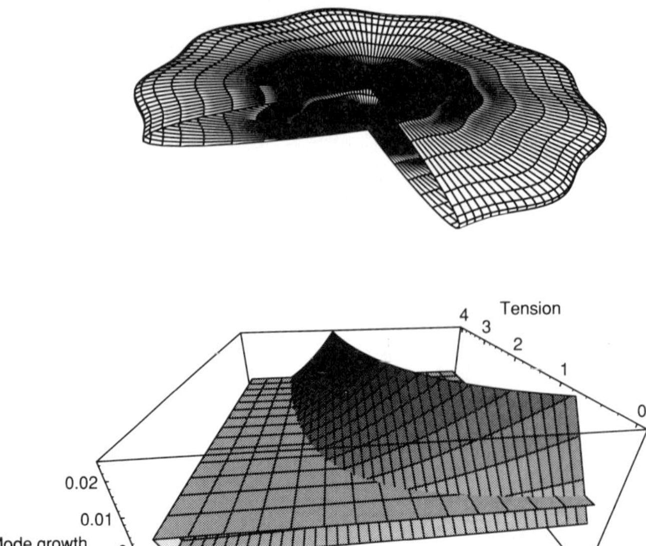

FIGURE 3

Dispersion relation for proteins diffusing on a membrane of varying tension. The membrane is parametrized as a cylinder of constant elliptic cross section with principle axes of 1000 nm and 100 nm. The mechanical instability initiating vesicle formation around the rim occurs at a wavelength corresponding to a vesicle of 50 nm in radius. Membrane tension (in units of pN/nm) is plotted along the axis going into the page. Parameters: $r_p = 2.5\,\mathrm{nm}$, $\kappa_p = .2\,\mathrm{nm}^{-1}$, $\mathcal{B} = 100\,\mathrm{pN}\cdot\mathrm{nm}$, $k_B T = 4.1\,\mathrm{pN}\cdot\mathrm{nm}$, $\zeta = 0.07\,\mathrm{pN}\cdot\mathrm{sec}\cdot\mathrm{nm}^{-3}$, $K = 0.0005\,\mathrm{pN}\cdot\mathrm{nm}^{-1}$, $C_o = 0.1\,\mathrm{nm}^{-1}$

References

1. J. Abney, *Protein organization and mobility in crowded biological membranes.*, Biophys. J. **64** (1993), A340.
2. R. Anderson and J. Kaplan, *Receptor-mediated endocytosis.*, Modern Cell Biol. **1** (1992), 1–52.
3. P. Burn, Fundamentals of Medical Cell Biology. Membranology and subcellular organelles. Cell Biol., 21–66, JAI Press., 1992, pp. 21–66, In: Membrane-protein-lipid interactions.
4. Y.-T. Chen, C. Holcomb, and H.-P. Moore, *Expression and localization of two small-molecular weight gtp-binding proteins, rab8 and rab10, by epitope tag.*, PNAS **in press** (1993).
5. R. DeBiasio, L.-L. Wang, G.W. Fisher, and D.L. Taylor, *The dynamic distribution of fluorescent analogues of actin and myosin in protrusions at the leading edge of migrating swiss 3t3 fibroblasts.*, The Journal of Cell Biology. **107** (1988), 2631–2645.
6. J.M. Fernandez, M. Vilalon, and P. Verdugo, *Reversible condensation of mast cell secretory products in vitro.*, Biophys. J. **59** (1991), 1022–1027.
7. T. Fujimoto, *Calcium pump of the plasma membrane is localized in caveloae.*, J. Cell Biol. **120** (1993), 1147–1157.
8. H.J. Geuze, G.J. Strous, H.F. Lodish, and A.L. Schwartz, *Intracellular site of asialoglycoprotein receptor-ligand uncoupling: double-label immunoelectron microscopy during receptor-mediated endocytosis.*, Cell **32(1)** (1983), 277–87.
9. W. Helfrich and M. Kozlov, *Bending tensions and the bending rigidity of fluid membranes.*, J. Physique II. **3** (1993), 287–92.
10. J. Heuser, Z. Q., and M. Clarke, *Proton pumps populate the contractile vacuoles of dictyostelium amoebae.*, J. Cell Biol. **121** (1993), 1311–1327.
11. N. Hooper, *More than just a membrane anchor*, Curr. Biol. **2** (1992), 617–619.
12. T.E. Kreis, *Regulation of vesicular and tubular membrane traffic of the golgi complex by coat proteins.*, Curr. Opin. Cell Biol. **4** (1992), 609–15.
13. S. Leibler, *Curvature instability in membranes*, Journal de Physique **47** (1986), 507–516.
14. R. Lipowsky, *Domain-induced budding of fluid membranes.*, Biophys. J. **64** (1993), 1133–1138.
15. V. Markin, *Lateral organization of membranes and cell shapes.*, Biophys. J. **36** (1981), 1–19.
16. A.P. Minton, *Lateral diffusion of membrane proteins in protein-rich membranes. a simple hard particle model for concentration dependence of the two-dimensional diffusion coefficient.*, Biophys. J. **55** (1989), 805–808.
17. R. Mortara and G. Koch, *An association between actin and nucleocapsid polypeptides in isolated murine retroviral particles.*, J. Submicro. Cytol. Pathol. **21** (1989), 295–306.
18. O.G. Mouritsen and M. Bloom, *Mattress model of lipid-protein interactions*

in membranes, Biophys. J. **46** (1984), 141–153.

19. G. Oster, L. Cheng, H.-P.H. Moore, and A. Perelson, *Vesicle formation in the golgi apparatus.*, J. Theo. Biol. **141** (1989), 463–504.
20. G. Oster and H.-P.H. Moore, Cell to Cell Signalling: From Experiments to Theoretical Models., 171–187, New York: Academic Press, 1989, pp. 171–187, In: The budding of membranes.
21. G Oster and A. Perelson, Cell Behavior: Shape, Adhesion and Motility., 35–54, 1988, pp. 35–54, In: The physics of cell motility.
22. G. Oster and A. Perelson, Frontiers in Mathematical Biology., Berlin: Springer-Verlag., 1993, In: Cell protrusions.
23. I. Sandoval, J. Bonifacino, R. Klausner, M. Henkart, and J. Wehland, *Role of microtubules in the organization and localization of the golgi apparatus.*, J. Cell Biol. **99** (1984), 113s–118s.
24. U. Seifert, K. Berndl, and R. Lipowsky, *Shape transformations of vesicles: phase diagram for spontaneous- curvature and bilayer-coupling models.*, Phys. Rev. A. **44** (1991), 1182–202.
25. A. Shariff and E.J. Luna, *Diacylglycerol stimulated formation of actin nucleation sites at plasma membrane*, Science **256** (1984), 245–247.
26. M.M. Sperotto, J.H. Ipsen, and O.G. Mouritsen, *Theory of protein induced lateral phase separation in lipid membranes*, Science **14** (1989), 79–95.
27. R. Szilard, *Theory and analysis of plates: Classical and numerical methods.*, Englewood Cliffs, N.J.: Prentice-Hall., 1974.
28. A. Tartakoff and J. Turner, Fundamentals of Medical Cell Biology. Membranology and subcellular organelles., 283–304, Greenwich, Conn.: JAI Press ., 1992, pp. 283–304, In: The Golgi apparatus.
29. P. Verdugo, *Mucin exocytosis.*, Amer. Rev. Respir. Dis. **144** (1991), 533–37.
30. J. Wehland, M. Henkart, R. Klausner, and I. Sandoval, *Role of microtubules in the distribution of the golgi apparatus: effect of taxol and microinjected anti-alpha-tubulin antibodies.*, Acad. Sci. USA. **80** (1983), 4286– 90.
31. G. Weiss, *First passage time problems in chemical physics.*, Adv. Chem. Phys. **13** (1967), 1–18.

DEPARTMENT OF MOLECULAR & CELL BIOLOGY, UNIVERSITY OF CALIFORNIA, BERKELEY, CA 94720

E-mail address: odelle@nature.berkeley.edu, goster@nature.berkeley.edu

Lectures on Mathematics in the Life Sciences
Volume **24**, 1994

Measurements of Molecular Transport in Small Systems

ELLIOT L. ELSON and HONG QIAN

ABSTRACT. Three methods based on optical microscopy are available for measuring the lateral transport of molecules in small systems such as individual living cells. Two of these, Fluorescence Correlation Spectroscopy (FCS) and Fluorescence Photobleaching Recovery (FPR), measure changes in the number ("occupation number") of fluorescence labelled molecules in a small open subregion of the sample. In an FCS measurement, the system rests in equilibrium while the occupation number undergoes microscopic spontaneous fluctuations. In an FPR measurement, a macroscopic concentration gradient is generated by photobleaching a fraction of the fluorophores in the observation region. A third method, Single Particle Tracking (SPT), determines the trajectory of individual particles visible in the light microscope. FCS and SPT measure microscopic changes and so require that an extensive record of occupation number fluctuations or particle trajectories be analyzed statistically to provide accurate measurements of diffusion coefficients. FPR measurements, based on macroscopic concentration changes, can yield valid diffusion coefficient values from a single recovery transient. In this paper we describe these three methods, discuss the validity of the measurements, and illustrate applications.

1. Introduction

Measurements of molecular transport in biological systems can provide important information about both physiological mechanisms and physical in-

1991 Mathematics Subject Classification, Primary: 92C05
Supported by NIH Grant GM 38838
This paper is in final form and no version of it will be submitted for publication elsewhere.

teractions that drive and constrain molecular motions. For example, according to a simple model of membrane organization, cell surface proteins are embedded in a fluid lipid bilayer and should diffuse freely, limited mainly by the viscosity of the bilayer (Singer and Nicolson, 1972). It has also been proposed that systematic rearward transport of cell surface particles is driven by a current of lipid that sweeps along with it slowly diffusing membrane protein aggregates (Bretscher, 1984). As described below, measurements of diffusion and systematic transport of cell surface proteins have shown that these models cannot account for the transport properties actually observed.

Methods for measuring the rates of random lateral diffusion and of systematic drifts or flows of proteins in individual cells have been based on optical microscopy. One approach is to measure the number of fluorescence labelled molecules in a small open subregion of the system. The number of molecules (the "occupation number") can change due to either systematic motion or diffusion. The rate of change of this number depends on the size of the observed subregion and the rate of molecular motion. Therefore, once the size of the observation region is known, measurements of the rate of change of the occupation number yield the rates of molecular motion, e.g., diffusion coefficients and drift velocities. Two methods have been developed based on this idea (Elson, 1985). One, called "Fluorescence Correlation Spectroscopy" or FCS, measures the spontaneous fluctuations of the occupation number about its constant mean value as the system rests in thermodynamic equilibrium or in a steady state condition of diffusion and flow. The other, called "Fluorescence Photobleaching Recovery" or FPR[1], measures the rate of relaxation of a macroscopic concentration gradient generated by an intense pulse of excitation light that irreversibly photolyzes a fraction of the fluorescent molecules in the observation region. In FCS many spontaneous fluctuations must be analyzed statistically to determine the desired diffusion coefficients and velocities of transport. In FPR a single macroscopic transient is sufficient in principle to yield these quantities.

A different approach tracks the position of one or a small cluster of molecules linked to a small particle that is visible by video enhanced contrast optical microscopy (Geerts et al., 1987) and that can be localized with high precision (Gelles et al., 1988). In this "Single Particle Tracking" or SPT method, diffusion coefficients and rates of systematic transport are determined from a statistical analysis of the trajectories of individual particles (Gross and Webb, 1988; Sheetz et al., 1989; Qian et al., 1991). Like FCS, the SPT method does not require a perturbation of the system to establish a macroscopic concentration gradient but does require the analysis of an extensive record of particle motion.

To use these methods with confidence it is necessary to assess the accuracy of the results they yield. For the occupation number methods there are uncertainties due to the random emission of fluorescence photons, i.e., "shot" noise. In SPT the analogous source of uncertainty is in the determination of the position of the observed particle tag. For both SPT and FCS there is a

[1] This method is also called "Fluorescence Recovery After Photobleaching" or FRAP.

more intrinsic uncertainty due to the stochastic character of random diffusion. Even if the time course and amplitude of a single spontaneous occupation number fluctuation could be measured with infinite accuracy, this would not accurately determine the phenomenological diffusion coefficient (Onsager, 1931). Rather many fluctuations must be analyzed statistically. Similarly, even if the position of a particle could be determined with infinite accuracy, a short record of its trajectory is insufficient to determine its diffusion coefficient. The statistical analyses are most conveniently accomplished in terms of temporal correlation functions computed from the measurements. In FCS, the diffusion coefficients are derived from a fluorescence fluctuation autocorrelation function, which corresponds to an occupation number autocorrelation function and which is computed from a time sequence of fluorescence intensity measurements (Elson and Magde, 1974). In SPT, diffusion coefficients are calculated from the mean square displacement of the observed particle, which is closely related to the correlation function of particle positions (Qian et al., 1991). Because they are based on the relaxation of macroscopic concentration gradients, FPR measurements are not influenced by these kinds of stochastic uncertainties.

2. Occupation Number Measurements

The fluorescence signal $f(t)$ measured in FPR and FCS is related to the concentration of fluorophores $c(\mathbf{r},t)$ at position $\mathbf{r}$ and time t and to the excitation intensity profile $I(\mathbf{r})$ as

$$f(t) = Q\int I(\mathbf{r})c(\mathbf{r},t)\ d^n\mathbf{r}$$

where $n(=1,2,3)$ is the dimensionality of the system. In an FPR experiment the time course of relaxation to the initial equilibrium concentration $<c>$ is

$$\Delta f(t) = Q\int I(\mathbf{r})\ \Delta c(\mathbf{r},t)\ d^n\mathbf{r}$$

where $\Delta f(t) = f(t) - <f>$ and $\Delta c(\mathbf{r},t) = c(\mathbf{r},t) - <c>$ represent macroscopic deviations from the equilibrium values and Q takes account of the absorption and emission characteristics of the fluorophore and the optical excitation and losses in the measurement system. In an FCS experiment the statistical analysis of the fluorescence fluctuations is carried out in terms of a fluorescence fluctuation autocorrelation function $G(\tau)$ where

$$\begin{aligned} G(\tau) &= \lim_{T\to\infty}(1/T)\int_0^T \delta f(t)\delta f(t+\tau)dt \\ &= Q^2\int I(\mathbf{r})I(\mathbf{r}') < \delta c(\mathbf{r},0)\delta c(\mathbf{r}',\tau) > d^n\mathbf{r}d^n\mathbf{r}' \end{aligned}$$

Here δ represents a microscopic fluctuation, $< \ldots >$ denotes a time or ensemble average, and we have supposed that the system is stationary. The phenomenological equation that governs concentration change due to diffusion with diffusion coefficient D and drift with velocity V along the x-direction is

$$\partial\delta c(\mathbf{r},t)/\partial t = D\nabla^2\delta c(\mathbf{r},t) - V\partial\delta c(\mathbf{r},t)/\partial x$$

To describe multicomponent mixtures, this equation is readily extended into a system of equations that can account, in addition, for chemical reactions among the components. The analysis of these equations for the interpretation of FPR and FCS measurements has been discussed in detail (Elson, 1985).

Due to the random character of fluorescence emission, the number of photons emitted by a fluorophore will fluctuate about its mean value according to a Poisson distribution. Hence, the accuracy of both FCS and FPR measurements will be influenced by the magnitude of these "shot noise" fluctuations. The magnitude of the shot noise will be determined by the duration, ΔT, of the individual intensity measurements and the mean number of photons produced per unit time, $\lambda I(\mathbf{r})$. Denoting $x(\mathbf{r}) = \lambda I(\mathbf{r})\Delta T$, the probability that the photocount p measured from a fluorophore at $\mathbf{r}$ in an interval ΔT should equal the number k is Prob $\{p = k\} = x(\mathbf{r})^k \exp(-x(\mathbf{r}))/k!$. More generally, if the probability density that the fluorophore is at $\mathbf{r}$ is $P(\mathbf{r})$, then

$$\text{Prob}\,\{p = k\} = \int [x(\mathbf{r})^k \exp(-x(\mathbf{r}))/k!]P(\mathbf{r})d^n\mathbf{r}$$

Suppose that the ideal, shot noise free fluorescence intensity is $\Phi(\mathbf{r}) = \lambda I(\mathbf{r})\Delta T$. Since $P(\mathbf{r})dr = \text{Prob}\{\Phi = x\}dx$, we have

$$\text{Prob}\{p = k\} = \int [x(\mathbf{r})^k \exp(-x(\mathbf{r}))/k!]\,\text{Prob}\,\{\Phi = x(\mathbf{r})\}d^n\mathbf{r}$$

This illustrates that the probability distribution of the detected photocounts p is a Poisson transformation of the distribution of Φ, independent of the detailed form of Prob $\{\Phi = x\}$ (Qian, 1990a). In principle, it is possible to carry out an inverse Poisson transformation to filter out the shot noise contribution (Qian, 1990b). It is also interesting to note that the Poisson transformation is a special example of Hidden Markov Models, which have recently received increasing attention in studies of signal processing (Rabiner and Juang, 1986).

To account for the statistical uncertainty in FCS measurements, we begin by noting that the fluorescence intensity, P, from a system of M particles is simply the sum of the intensities measured from the individual particles, $P = \sum p_i$. If we measure P at a series of N times, then the mean fluorescence signal is $S_1 = \sum P_i/N$. In an FCS experiment we calculate

$$S_2(m) = \sum_{k=1}^{N}(P_k P_{k+m})/N - \left[\sum_{k=1}^{N} P_k/N\right]^2$$

to yield the correlation function $< S_2(m) >= G(m\Delta T) =< P(0)P(m\Delta T) > - < S_1 >^2$, where ΔT is the interval between measurements. The statistical uncertainty is represented by the variance of S_2, $< (\Delta S_2(m))^2 >$. Then the signal to noise ratio, Ξ, is $\Xi =< S_2(m) > /[< (\Delta S_2(m))^2 >]^{\frac{1}{2}}$. Expressions for $< S_2(m) >$ and $< (\Delta S_2(m))^2 >$ have been presented (Qian, 1990a). Analysis of the signal to noise ratio is complex and leads to two limiting cases (Qian, 1990a):

(1) Under conditions in which q, the measured intensity per fluorophore, is high and m, the number of fluorophores in the observation region, is large, Gaussian occupation number statistics prevail, and the contribution from shot noise is small; then Ξ is independent of m and q and is limited only by the stochastic nature of the fluctuation measurements and is governed by N, the number of **independent** intensity measurements. (Note that successive intensity measurements will not be independent if, as is usually true, the dwell time ΔT is less than the correlation time, τ, for the observed process, e.g., diffusion across the observation region. When $\Delta T < \tau$, Ξ varies as $(T/\tau)^{-1/2}$, i.e., as (number of independent measurements)$^{-1/2}$ [cf. Qian et al., 1991]).

(2) If m is small, Poisson rather than Gaussian statistics prevail, and Ξ will depend also on m. If q and m are both small, the contribution of shot noise is also significant. Then Ξ depends on both q and m as well as N. If q is large, but m is small, the shot noise contribution is small, and Ξ depends only on m and N.

3. Single Particle Tracking

SPT measurements provide the trajectory of a particle, $\mathbf{r}(t) = [x(t),\ y(t),\ z(t)]$, from a sequence of position measurements. Values for diffusion coefficients and drift velocities are conveniently obtained from the mean squared displacement of the particle, $\rho(t)$, as a function of time:

$$\text{(1)} \qquad \rho(t) =< |\mathbf{r}(t) - \mathbf{r}(0)|^2 >= \int\int P(\mathbf{r}')|\mathbf{r} - \mathbf{r}'|^2 P(\mathbf{r}|\mathbf{r}', t) d^n\mathbf{r} d^n\mathbf{r}'$$

where $P(\mathbf{r})$ is the steady-state distribution of particle position and $P(\mathbf{r}|\mathbf{r}', t)$ is the probability that a particle originally at $\mathbf{r}'$ will be at $\mathbf{r}$ after a time period t. Values of $\rho(t)$ are calculated from particle positions measured sequentially at time intervals ΔT: e.g.,

$$\rho_x(n\Delta T) = \sum_{i=0}^{N} (x_{i+n} - x_i)^2/(N+1) \ .$$

If the particle is both randomly diffusing with diffusion coefficient D and drifting with velocity $\mathbf{V}$, then in a two-dimensional system appropriate for analysis of diffusion on a cell membrane, $P(\mathbf{r}|\mathbf{r}', t) = (1/4\pi Dt)\exp[-|\mathbf{r} - \mathbf{r}' - \mathbf{V}t|^2/4Dt]$ and $\rho(t) = 4Dt + |\mathbf{V}t|^2$. Hence, the contributions of drift and diffusion are readily discerned and quantitatively characterized in a plot of $\rho(t)$ versus t (Sheetz et al., 1989).

As with FCS, even with perfectly precise measurements of particle positions, the values of $\rho(t)$ calculated from experimental data will have statistical variances due to the stochastic character of diffusion. This variance will diminish as the number of position measurements increases. To assess the validity of experimentally determined diffusion coefficients, it is essential to determine the dependence of the variance on the number of measurements. We first consider independent measurements of the squared displacement at time t, $\xi_t = |\mathbf{r}(t) - \mathbf{r}(0)|^2$. The probability distribution of ξ_t considered as a random variable is $\text{Prob}[z \leq \xi_t \leq z + dz] = (1/4Dt) \exp(-z/4Dt)dz$. Suppose that we have K independent measurements $\xi_t(i)$ $(i = 1, ..., K)$. Because $\xi_t(i)$ is a random variable, the sum of independent measurements of ξ_t and therefore the mean $\rho(t) = [\xi_t(1) + \xi_t(2) + \cdots + \xi_t(K)]/K$ are also random variables. The probability distribution of ρ can then be derived as successive convolutions of $\text{Prob}[z \leq \xi_t \leq z + dz]$ with itself to yield $\text{Prob}[z \leq \rho(t) \leq z + dz] = [(1/4Dt)^K \exp(-zK/4Dt) K^K z^{K-1}/(K-1)!]dz$ (Qian et al., 1991). Using this one may show that the relative error in the determination of ξ is $(< (\Delta\rho)^2 > / < \rho >^2)^{1/2} = K^{-1/2}$.

In actual SPT measurements the value of $\rho(t)$ is determined from a sequence of N *consecutive* positions $\mathbf{r}$ at time intervals of ΔT. Then $\rho_n = \rho(n\Delta T)$ is obtained by averaging over $N - n + 1$ measurements, and so the successive determinations of $\rho(t)$ are not independent, due to the overlap between the measurements. This leads to an increase in the relative error depending on n. Calculation of the relative error is complex (Qian et al., 1991). When $N \gg n$, $[< (\Delta\rho_n)^2 > / < \rho_n >^2]^{1/2} = [(2n^2 + 1)/3n(N - n + 1)]^{1/2} \approx [2n/3(N-n+1)]^{1/2} \approx [2n/3N]^{1/2}$. Hence, the variance increases with increasing n because the larger n, the smaller the number of statistically independent samples of displacement within the interval $N\Delta T$. This analysis is extended to account for measurements of the velocity of systematic drift by Qian et al., (1991).

Uncertainties in measurements of the positions of particles are typically small and vary randomly from one video frame to the next (Gelles, 1988). The contribution of this type of uncertainty is analogous to shot noise; it should appear in $\rho(t)$ only at $t = 0$. Therefore, the major uncertainty in the measurement of diffusion and drift by SPT arises from the stochastic character of random particle motion (Qian et al., 1991).

4. Relationship of the Methods

FCS, FPR, and SPT can all be interpreted in terms of an elementary theory of molecular transport represented in terms of a stochastic (Markov and Gaussian) process with the transition probability distribution $P(\mathbf{r}|\mathbf{r}', t)$. In SPT measurements the positions of a particle are weighted equally in calculating the mean square displacement $\rho(t)$. As indicated above [Eq. (1)], the mean square displacement of a particle can be expressed in terms of $P(\mathbf{r}|\mathbf{r}', t)$. It is also straightforward to show that for motion within a finite space there is a close connection between the positional correlation function of a particle, $g_r(t) =< r(t)r(0) >$, and its mean square displacement (Qian et al., 1991): $g_r(t) =< r^2 > +\rho(t)/2$.

In contrast to SPT, in FCS and FPR measurements the position of the particle is weighted by the excitation intensity profile, $I(\mathbf{r})$, according to its location relative to the center of the excitation laser beam and relative to the focal plane of the microscope. For example, the fluorescence fluctuation autocorrelation function can be represented as

$$G(t) = P^2 \int\int I(\mathbf{r})I(\mathbf{r}')P(\mathbf{r}')P(\mathbf{r}|\mathbf{r}', \tau)d^n\mathbf{r}d^n\mathbf{r}'$$

in which the position of the fluorophores at positions $\mathbf{r}$ and $\mathbf{r}'$ are weighted by the excitation light intensities $I(\mathbf{r})$ and $I(\mathbf{r}')$. The distribution of the excitation light intensity in FCS and FPR measurements imposes a characteristic distance on the measurement. In the simplest spot photobleaching measurements, that characteristic distance is the radius, w, of the Gaussian excitation laser light intensity: $I(r) = I_0 \exp(-2r^2/w^2)$. Then, simple diffusion with diffusion coefficient D has a characteristic time, τ_D, set by this distance, $\tau_D = w^2/4D$, which plays an important role both in FCS and FPR measurements of diffusion (Elson, 1985). Similarly, in dynamic light scattering (DLS) the weighting function is $I(\mathbf{r}) = \exp(i\mathbf{r} \cdot \mathbf{q})$ where $\mathbf{q}$ is the scattering vector and $1/|q|$ is the characteristic distance imposed on the diffusion measurement (Cummins et al., 1969; Qian et al., 1991). Analogous to the characteristic time $w^2/4D$ for FCS, the charactersitic time for DLS is $1/Dq^2$. The relationship between characteristic time and distance in FPR and FCS can be used to accommodate measurements to experimentally accessible ranges. For example, in measurements of very fast diffusion, the value of w can be increased to bring the τ_D into a range convenient for measurement.

In contrast to FPR, FCS, and DLS, there is no characteristic distance in an SPT measurement. To measure diffusion with high relative precision it is merely necessary to allow the particle to move to a large mean square displacement. Of course, systematic errors in the measurement could impose additional limits.

5. Discussion and Illustration

5.1. Two- and Three-Dimensional Systems. Application of these methods to quasi-two-dimensional systems such as cell membranes is straightforward because the plane of focus of the microscope can be made to coincide with the plane in which the particle motion is to be measured. Extension to three-dimensional systems can be difficult because the measured fluorescence intensity emitted from a particle or molecule diminishes the further it is from the focal plane. To account for the effects of this variation on FCS measurements it is necessary to analyze how the excitation intensity and the collected emission intensity vary with distance from the focal plane. This is accomplished in terms of the point spread function of the microscope, which relates the intensity of light transmitted to a point $\mathbf{r}'$ in the microscope from a position $\mathbf{r}$ in the sample. It is also necessary to take account of a field aperture placed before the photomultiplier that reduces the off-focus intensity collected. This problem has been discussed in detail for the confocal microscope optics typically used for FCS (Koppel et al., 1976; Qian and Elson, 1991). In SPT measurements differential interference contrast (DIC) or darkfield microscope optics are typically used with high levels of incident illumination. Therefore shot noise is usually not a problem. An advantage of DIC optics is a very narrow depth of focus so that particles remain visible only within a short distance of the focal plane. In SPT measurements the observed particle can therefore disappear by leaving the focal plane, and, even if it remains in view, its displacement from the focal plane is difficult to measure precisely.

5.2. Advantages and Disadvantages of the Methods. An advantage in principle of FCS is that the system rests unperturbed in equilibrium during the measurement. In an FPR measurement the system is perturbed from equilibrium by photobleaching in a localized area of the system. It is assumed in most analyses of FPR measurements that the photobleaching can be described as an irreversible process that is first order in the concentration of fluorophores and with a rate proportional to the incident light intensity. These assumptions are rarely tested and their failure could lead to errors in the evaluation of FPR measurements (Elson, 1985). For example, if the photobleaching process were reversible, some of the measured fluorescence recovery could be due to this process rather than to diffusion or drift of unbleached fluorophores into the observation region. FCS, however, has the disadvantage of requiring the measurement of a long sequence of microscopic fluorescence changes. This imposes a stringent requirement for stability on the measured system, which usually makes this method unsuitable for application to living systems. Any motion of a cell, ruffling or locomotion, for example, might produce a fluorescence change, which would overwhelm the small fluorescence changes due to occupation number fluctuations. In contrast FPR requires only a single macroscopic recovery transient to characterize transport properties within the limits of precision of the measurement. Major advantages of the SPT method are its high spatial resolution, which enables measurements of motions within very small domains, and the ability to detect motions exhibited by only a small fraction of the labelled molecules. A disadvantage of SPT in studies of the motion of cell surface molecules is that the marker beads are polyvalent for binding to the surface molecules. The number of molecules bound to the beads is difficult to determine and the polyvalent binding could itself influence the dynamic state of the bound molecules. The effective crosslinking of these molecules could activate dynamic cellular processes such as capping or other forms of systematic molecular transport. Another disadvantage of SPT is that, like FCS, an extensive record of random motion is required for a statistically accurate characterization of diffusion coefficients.

5.3. Examples:

5.3.1. FCS Measurements of the Motions of Beads in Actin Gels. Actin filaments form random viscoelastic networks that have an important role in determining cell shape and mechanical consistency and in driving cellular dynamic processes (Elson, 1988). One approach to characterizing the mechanical properties of actin gels is to measure the motion of particles embedded in gels reconstituted from filaments polymerized *in vitro* from purified actin. FPR and DLS have been used to study the mesh size and other microscopic characteristics of reconstituted actin filament networks (e.g., Luby-Phelps et al.,

1988; Hou et al., 1990a, 1990b; Newman et al., 1989; Seils et al., 1990; and Schmidt et al., 1989). In a recent study FCS was used to measure the motion of fluorescence labelled beads of various sizes in reconstituted actin gels (Qian et al., 1992). In these measurements the number of beads in the observation region was small, typically less than ten. FPR measurements on this system were difficult because the recovery curves varied substantially from one to another for statistical reasons, as described above. This variability made it difficult to discern the spatial and temporal properties of the gel that were of interest. By applying FCS instead of FPR, the large occupation number fluctuations due to the small number of observed particles is converted from a disadvantage to an advantage. In this study the FCS measurements indicated that the fluorescent beads were constrained in small cages formed by the gel and that the gel had a dynamic structure that varied over time and from one region of the gel to another. These measurements show the applicability of FCS to stable systems in which the motions of highly labelled particles present at low concentration yield large fluctuation signals.

5.3.2. Random Diffusion and Systematic Drift on Cell Surfaces. Systematic rearward transport of particles on locomoting cells has been attributed to a current of membrane lipid that was thought to sweep large particles or membrane protein aggregates along with it. An alternative hypothesis is that the rearward motion results from the operation of motor proteins that link surface molecules to the underlying cortical cytoskeleton and drag them toward the back of the cell. It is difficult to test the dynamic character of randomly diffusing and systematically transported molecules using FPR because many particles in both dynamic states contribute to the measured signal. In particular the motion of systematically transported particles is frequently difficult to discern against the background of randomly diffusing particles that are usually present in greater proportion. SPT provides a detailed characterization of the motion of both randomly diffusing and systematically transported individual particles. Studies of the motions of 40 nm particles linked to the surface of mouse bone-marrow macrophage by the lectin concanavalin A showed that the particles could undergo rapid reversible transitions between random diffusion and systematic transport and that the diffusion coefficient of particles being systematically transported was markedly reduced compared to the randomly diffusing particles (Sheetz et al., 1989). These results are difficult to reconcile with a lipid flow model, but are consistent with the concept that crosslinked cell surface proteins link reversibly to cytoskeletal motors and are drawn by them toward the back of the cell. SPT was also used to measure the motion of randomly diffusing particles on rapidly locomoting fish epidermal cells (Kucik et al., 1990). It was shown that the diffusing particles experienced no systematic drift relative to the cell carrying them and therefore that effects of lipid currents could not be detected in the motions of these particles.

These measurements show how SPT is especially useful in characterizing the motions of particles that can exist in and can undergo transitions among different dynamic states.

5.3.3. Voltage-Dependent Sodium Channels Exist in Different Dynamic States in Different Domains of Neurons. Voltage-dependent sodium channels are essential for the propagation of action potentials in nerve cells. It is therefore interesting to determine the distribution and mobility of these channels in different regions of neurons. The channels could be labelled with specific fluorescent tags derived from well characterized toxins. Using these specific labels it was possible to show that nearly all labelled channels diffuse freely and rapidly in the membrane of the cell body (Angelides et al., 1988). In contrast, on the axon hillock (the initial portion of the axon as it emerges from the cell body) the lateral diffusion of the channels is restricted. On this region of the cell the channels diffuse much more slowly. Moreover, there is a barrier that prevents diffusion of channels between the hillock and the cell body. This is a pattern of organization specific to the voltage-dependent sodium channel and is not shown either by a lipid probe or by a relatively nonspecific lectin label.

FPR and other fluorescence microscopy methods have been used to characterize the development and organization of the cell membrane into different domains in which, in addition to voltage-dependent channels, molecules such as acetylcholine receptors and specific sperm antigens are either constrained or are free to diffuse randomly (e.g., Dubinsky et al; 1989; Cowan et al., 1987; cf., Elson, 1993). The proper spatial and dynamic organization of these domains is presumably essential for the transmission of signals or for other interactions within and among cells.

5.3.4. Diffusion in Small Systems. Because of their high spatial resolution, methods such as FPR and SPT can be used to characterize small domains (e.g., Angelides et al., 1988; Yechiel and Edidin, 1987; Edidin et al., 1991; Vaz, 1992; Glaser, 1992; Edidin, 1992; Wolf, 1992). In measurements of the diffusion of molecules constrained to remain within domains, it is important to take account of the effects of the domain size on the apparent diffusion rate. Mathematically, this corresponds to considering the domain boundaries when solving the diffusion equation. This subject has been analyzed for FPR measurements (Qian and Elson, 1988; Elson and Qian, 1989). It was shown that when the fluorescent probe molecules are confined to a small area, both the size of the area and its shape can influence the apparent rate of recovery. As the area available for diffusion decreases, the apparent diffusion rate increases relative to that for an infinite sample area. This increase in small areas is due to the absence of the contributions from remote points in large domains that contribute relatively slow recovery components. Furthermore, as the sample area becomes more elongated, the diffusion rate appears slower than for compact sample domains with the same area. The apparent fraction of mobile molecules also depends on the area available for diffusion. If the area of the bleached spot is comparable to the available area, the photobleaching pulse could eliminate a substantial fraction of the fluorophores that can participate in recovery. This would cause the fraction of the fluorescence that recovered after photobleaching to be less than unity, even if all the fluorophore were mobile. These effects are relatively simple to account for (Qian and Elson, 1988; Elson and Qian, 1989) and can be useful in the interpretation of experiments, as, for example, in assessing the size of "cages" in actin gels from the enhancement in the rate of diffusion of fluorescent beads that they contain (Qian et al., 1992).

REFERENCES

Angelides, K.J., L.W. Elmer, D. Loftus, and E. Elson, Distribution and lateral mobility of voltage-dependent sodium channels in neurons. *J. Cell Biol.* **106** (1988), 1911-1925.

Bretscher, M.S., Endocytosis: relation to capping and cell locomotion, *Science* **224** (1984), 681-686.

Cowan, A.E., D.G. Myles, and D.E. Koppel, Lateral diffusion of the PH-2 protein on guinea pig sperm: Evidence that barriers to diffusion maintain plasma membrane domains in mammalian sperm, *J. Cell Biol.* **104** (1987), 917-923.

Cummins, H.Z., F.D. Carlson, T.J. Herbert, and G. Woods, Translational and rotational diffusion constants of tobacco mosaic virus from Rayleigh linewidths, *Biophys. J.* **9** (1969), 520-546.

Dubinsky, J.M., D.J. Loftus, G.D. Fischbach, and E.L. Elson, Formation of acetylcholine receptor clusters in chick myotubes: migration or new insertion? *J. Cell Biol.* **109** (1989), 1733-1743.

Edidin, M., S.C. Kuo, M.P. Sheetz, Lateral movements of membrane glycoproteins restricted by dynamic cytoplasmic barriers, *Science* **254** (1991), 1379-1382.

Edidin, M., The variety of cell surface membrane domains, *Comments Mol. Cell. Biophys.* **8** (1992), 73-82.

Elson, E.L. and D. Magde, Fluorescence correlation spectroscopy, I. Conceptual basis and theory, *Biopolymers* **13** (1974), 1-27.

Elson, E.L. and H. Qian, Interpretation of fluorescence correlation spectroscopy and photobleaching recovery in terms of molecular interactions, in *Fluorescence Microscopy of Living Cells in Culture, Part B*, D.L. Taylor and Y-L. Wang, eds., Academic Press, San Diego, CA, 1989, 307-332.

Elson, E.L., Cell Polarity: Barriers to diffusion, *Curr. Biol.* **3**, (1993), 152-154.

Elson, E.L., Cellular mechanics as an indicator of cytoskeletal structure and function, *Ann. Rev. Biophys. Biophys. Chem.* **17** (1988), 397-430.

Elson, E.L., Fluorescence correlation spectroscopy and photobleaching recovery, *Ann. Rev. Phys. Chem.* **36** (1985), 379-406.

Geerts, H., M. DeBrabander, R. Nuydens, S. Geuens, M. Moeremans, J. De Mey, and P. Hollenbeck, Nanovid tracking: A new automatic method for the study of mobility in living cells based on colloidal gold and video microscopy, *Biophys. J.* **52** (1987), 775-785.

Gelles, J., B.J. Schnapp, and M.P. Sheetz, Tracking kinesin-driven movements with nanometer-scale precision, *Nature (Lond.)* **331** (1988), 450-453.

Glaser, M., Characterization and formation of lipid domains in vesicles and erythrocyte membranes, *Comments Mol. Cell. Biophys.* **8** (1992), 37-51.

Gross, D.J. and W.W. Webb, Cell surface clustering and mobility of the liganded LDL receptor measured by digital fluorescence microscopy, in *Spectroscopic Membrane Probes. Vol II*, L.M. Loew, ed., CRC Press, Boca Raton, FL, 1988, pp. 19-48.

Hou, L., F. Lanni, and K. Luby-Phelps, Tracer diffusion of F-actin and ficoll mixtures: Toward a model for cytoplasm, *Biophys. J.* **58** (1990b), 31-43.

Hou, L., K. Luby-Phelps, and F. Lanni, Brownian motion of inert tracer macromolecules in polymerized and spontaneously bundled mixtures of actin and filamin, *J. Cell Biol.* **110** (1990a), 1645-1654.

Koppel, D.E., D. Axelrod, J. Schlessinger, E.L. Elson, and W.W. Webb, Dynamics of fluorescence marker concentration as a probe of mobility, *Biophys. J.* **16** (1976) 1315-1329.

Kucik, D.F., E.L. Elson, and M.P. Sheetz, Cell migration does not produce mebrane flow, *J. Cell Biol.* **111** (1990), 1617-1622.

Luby-Phelps, K., F. Lanni, and D.L. Taylor, The submicroscopic properties of cytoplasm as a determinant of cellular function, *Ann. Rev. Biophys. Biophys. Chem.* **17** (1988), 369- 396.

Newman, J., N. Mroczka, and K.L. Schick, Dynamic light scattering measurements of the diffusion of probes in filamentous actin solutions, *Biopolymers* **28** (1989), 655- 666.

Onsager, L, Reciprocal relations in irreversible processes, I, *Phys. Rev.* **37** (1931), 405-426.

Qian, H. and E.L. Elson, Analysis of confocal laser-microscope optics for 3-D fluorescence correlation spectroscopy, *Appl. Optics* **30** (1991), 1185-1195.

Qian, H. and E.L. Elson, Measurement of diffusion in closed region by FPR, *J. Cell Biol.* **106** (1988), 1921-1923.

Qian, H., Inverse Poisson transformation and shot noise filtering, *Rev. Sci. Instrum.* **61** (1990), 2088-2091.

Qian, H., On the statistics of fluorescence correlation spectroscopy, *Biophys. Chem.* **38** (1990), 49-57.

Qian, H., E.L. Elson, and C. Frieden, Studies on the structure of actin gels using time correlation spectroscopy of fluorescent beads, *Biophys. J.* **63** (1992), 1000-1010.

Qian, H., M.P. Sheetz, and E.L. Elson, Single particle tracking. Analysis of diffusion and flow in two-dimensional systems, *Biophys. J.* **60** (1991), 910-921.

Rabiner, L.R. and B.H. Juang, An introduction to hidden Markov models, *IEEE ASSP Magazine*, (January 1986), pp. 4-16.

Schmidt, C.F., M. Barmann, G. Isenberg, and E. Sackmann, Chain dynamics, mesh size, and diffusive transport in networks of polymerized actin. A quasielastic light scattering and microfluorescence study, *Macromolecules* **22** (1989), 3638- 3649.

Seils, J., B.M. Jockusch, and T.H. Dorfmuller, Dynamics of F-actin and F-acin/filamin networks as studied by photon correlation spectroscopy, *Biopolymers* **30** (1990), 677-689.

Sheetz, M.P., S. Turney, H. Qian, and E.L. Elson, Nanometre-level analysis demonstrates that lipid flow does not drive membrane glycoprotein movements, *Nature (Lond.)* **340** (1989), 284-288.

Singer, S.J. and G.L. Nicolson, The fluid mosaic model of the structure of cell membranes, *Science* **175** (1972), 720-731.

Tocanne, J.F., Detection of lipid domains in biological membranes, *Comments Mol. Cell. Biophys.* **8** (1992), 53-72.

Vaz, W.L., Translational diffusion in phase-separated lipid bilayer membranes, *Comments Mol. Cell. Biophys.* **8** (1992), 17-36.

Wolf, D.E., Lipid domains: The parable of the blind men and the elephant, *Comments Mol. Cell. Biophys.* **8** (1992), 83-95.

Yechiel, E. and M. Edidin, Micrometer-scale domains in fibroblast plasma membranes, *J. Cell Biol.* **105** (1987), 755-760.

ELLIOT L. ELSON
DEPARTMENT OF BIOCHEMISTRY AND MOLECULAR BIOPHYSICS
DIVISION OF BIOLOGY AND BIOMEDICAL SCIENCES
WASHINGTON UNIVERSITY SCHOOL OF MEDICINE
ST. LOUIS, MISSOURI 63110 U.S.A.
ELSON_E@BIOCHM.WUSTL.EDU

HONG QIAN
PHYSICS OF COMPUTATION LABORATORY
DIVISION OF CHEMISTRY
CALIFORNIA INSTITUTE OF TECHNOLOGY
PASADENA, CALIFORNIA 91125 U.S.A.
HONG@HOPE.CALTECH.EDU

Lectures on Mathematics in the Life Sciences
Volume **24**, 1994

On Peeling an Adherent Cell from a Surface

MICAH DEMBO

ABSTRACT. Biological adhesion is frequently mediated by specific membrane proteins (adhesion molecules). We present a simple model of the molecular physics of adhesion molecules in the moving boundary layer that separates a region of adherent membrane and nonadherent membrane. The model consists of combining the equations for deformation of an elastic membrane with equations for the chemical kinetics of the adhesion molecules. Mechanochemical coupling occurs through a set of constitutive laws relating bond stress and strain to the chemical rate constants governing formation and rupture of the adhesive bonds in the transition layer. Depending on the qualitative form of these constitutive laws, adhesion molecules can be classified as either "slip" bonds (stress increases the rate constant for bond rupture), "catch" bonds (stress decreases the rate constant for bond rupture) or "ideal" bonds (stress has no effect on the rate constant for bond rupture). In the case of catch bonds we show that adhesion cannot be reversed by simple application of tension; whereas in the case of slip bonds and ideal bonds, reversible peeling and attachment is possible. In the case of ideal bonds we derive the analytic theory for the peeling velocity, and we extend this theory to obtain an equation for the rolling velocity of granulocytes in the presence of fluid flow.

1. Introduction

Because of the natural tendency to focus on the positive aspect of adhesion, it is easy to forget the underlying fact that cells in general are not adhesive for each other and that at root there is usually a very strong and long ranged

---

1991 Mathematics Subject Classification, Primary: 92C10.
Supported by NIAID Grant RO1-AI21002.
This paper is in final form and no version of it will be submitted for publication elsewhere.

repulsive potential between biological surfaces. The physical basis of this repulsive potential (Bongrand and Bell, 1984; Evans, 1993) is to be found in the ubiquitous polysaccharide layer that coats the exterior surface of cell membranes, i.e., the glycocalyx. When a cell membrane comes into contact with an inert surface or with another cell, the polysaccharide chains of the glycocalyx are compressed and lose conformational entropy. Simultaneously, water is excluded from the vicinity of the hydrophilic sugar moeities thereby increasing the chemical potential of the solvent. If this were not enough, the negative charges on terminal sialic acids are brought into closer proximity thereby increasing the overall electrostatic potential (see Figure 1). The resulting repulsion is complex to deduce from first principles but is easy to understand intuitively. It is well documented in simple model systems and in one form or another is responsible for common phenomena such as the so-called "steric stabilization" of colloid suspensions and also the action of ionic detergents.

The biological basis of the repulsive potential is also easy to understand when one realizes that, at least on the cellular level, the first step in both predation and parasitism is the adhesion of a foreign surface to the plasma membrane. Very early in evolution cells apparently learned to maintain a slippery surface and to distrust all contact. Their success at mastering this lesson can be gauged from the fact that as of our present epoch there are no "general" cellular pathogens. For example, the known viruses each have an exquisitely restricted host range, and in fact viruses survive only by employing elaborately tailored strategies to breach the glycoalyx of their intended victims.

Given that adhesion is a dangerous and tricky business undertaken only for good and compelling physiological purposes, we should not be surprised to find that evolution has met each such contingency by yielding a specific class of macromolecule; i.e., the cell adhesion molecules or CAMs. We henceforth adopt the thesis that the *raison d'etre* of the various adhesion molecules is precisely to overcome the repulsive barrier of the glycoalyx when and only when it is necessary for a specific physiological function.

The known CAMs of mammalian tissue cells are all integral membrane proteins, and although the data are by no means complete, it is common to classify them as either integrins, selectins, ICAMs, or cadherins (see review by Springer, 1990). Integrins are heterodimers where both α and β subunits are involved in recognizing certain oligopeptides commonly found on proteins of the extracellular matrix. The binding affinity of the integrins depends on calcium, and the cytoplasmic portion interacts strongly with the cytoskeleton. The ICAMs are monomers or homodimers and, unlike the integrins, they frequently function to promote transient cell-to-cell adhesion necessary for signalling. They are a large and phylogenetically dispersed group that includes not only the immunoglobulins themselves but also homologous molecules that share the typical immunoglobulin domain (examples include the T-cell receptor, the MHC class I and class II, and the neural cell adhesion molecule or N-CAM).

The selectins are found on neutrophils, monocytes, lymphocytes, platelets, and endothelial cells. They function to regulate leukocyte binding to the endothelium at sites of inflammation. They are monomeric, their binding affinity is calcium-dependent, and they seem to recognize only carbohydrate (the N-terminal region of selectins is homologous to the carbohydrate binding domains on the animal lectins). Finally the cadherins (Takeichi, 1991) are a special class of monomeric adhesion molecules, all with very similar structure, that function in morphogenesis and in maintaining the structural integrity of solid tissue (e.g., uvomorulin or epithelelial cadherin, L-CAM or liver cell adhesion molecule, and desmoglein or desmosome cadherin). All cadherins are homophilic and their binding activity is strongly regulated by calcium. Cadherins are responsible for the tendency of mixed cell suspensions to sort themselves out into clumps of a uniform histological type.

The purpose in what follows is to provide some addition to our knowledge of biological adhesion by applying the perspective of physical and mathematical analysis. To achieve this we will present and analyze a theoretical model of membrane-to-substrate adhesion during attachment, at equilibrium, and during peeling. The model is not meant to be all inclusive: it is devised as a means to vitalize important quantitative concepts in a setting of minimum extraneous detail and complexity.

2. Some Basic Modeling Concepts

In this spirit consider an elastic membrane of total area A_{mem} and a flat rigid substrate of total area $A_{sub} \gg A_{mem}$. Let the surface of both the membrane and the substrate contain complementary adhesion molecules, uniformly distributed and firmly anchored (i.e., not free to diffuse laterally). Following convention, we refer to the molecules on the membrane as "receptors" and the molecules on the substrate as "sites."

We denote the total density of sites per unit area of substrate by n_{stot} and the total density of receptors per unit area of the membrane by n_{rtot}. We assume that the area modulus of the membrane is very large so that n_{rtot} is not affected by dilations. The densities of free or unbonded receptors and sites are given by n_r and n_s, respectively. It considerably simplifies the algebra if the number of sites is very much greater than the number of receptors $n_{stot} \gg n_{rtot}$, and we will henceforth assume that this is true. The density of adhesive bonds per unit area of membrane, n_b, is regarded as a continuous function of time and of position on the membrane manifold. We will assume that the adhesive bonds between the membrane and the substrate span the intervening gap along the shortest possible path; *i.e., the bonds are always perpendicular to the substrate.*

To avoid messy paradoxes it is important to remember that the notion of adhesive contact between a membrane and a substrate is not an absolute notion but is only defined relative to the properties of the bonds that mediate the contact. We therefore say that a point on the surface of the cell is in "contact" with the substrate at time t if $n_b > \delta_b$, where n_b is the local bond density and δ_b is a very small positive quantity called the contact threshold.

The set of all points such that $n_b > \delta_b$ is called the contact region or the contact zone; the measure of the contact region is called the contact area, A.

We view the dynamics of peeling and attachment of the membrane as dominated by events in a thin boundary layer straddling the edge of the contact region. Therefore, to study peeling at an arbitrary point on the circumference of the contact region, it is natural to adopt a "stretched" boundary layer coordinate as illustrated in Figure 1. In this coordinate system we neglect spatial variations unless they occur over distances on the order of the thickness of the boundary layer. In particular, because the boundary layer is very thin compared to the overall dimensions of the membrane, we can neglect variations of all quantities in the direction tangent to the circumference of the contact region. This maneuver and the related boundary layer expansion is closely related to the standard treatment of a propagating shock front. One may consult various texts on hydrodynamics for detailed discussions.

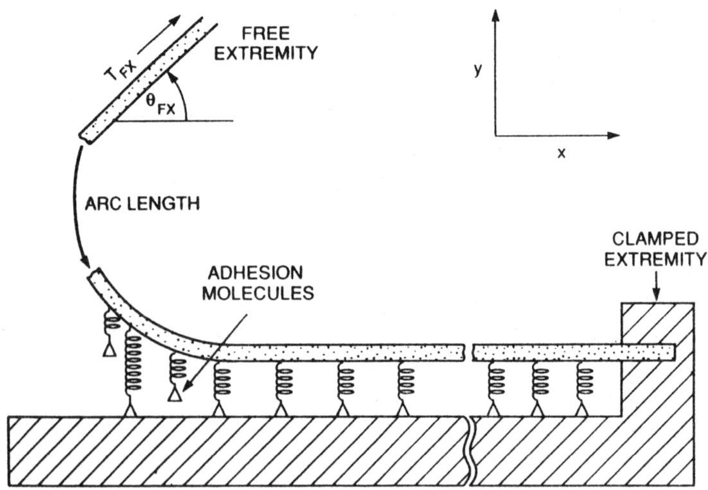

Figure 1. The geometry of membrane-to-surface adhesion in the boundary layer at the edge of the contact region. The surface to which the membrane binds is taken as coincident with the X-axis of Cartesian coordinates. Position along the contour of the membrane is tracked by the arc length coordinate, s. At the free extremity ($s \to -\infty$), a tension T_{fx} is applied to the membrane at a particular angle (θ_{fx}) with respect to the surface. At the contact extremity ($s \to +\infty$), the membrane is firmly atached to the surface by the adhesive bonds in the interior of the contact region. The origin of arc length is fixed with respect to the point where the bond density has dropped to a very small threshold level δ_b. Thus the boundary layer moves in response to changes in the bond density. The velocity of this motion is called the peeling velocity, V_{pl}.

Since tangential variations are neglected, the shape of the membrane in the neighborhood of a point on the edge of the contact region at time t can be described parametrically by a locus of points $[X(s,t), Y(s,t)]$, where s is arc length measured along a path in the plane of the membrane and normal to the boundary of the contact region. By convention, the origin of arc length is taken at the exact point where $n_b = \delta_b$. Thus the membrane is adherent for $s > 0$ and detached for $s < 0$. The so called "outer regions " as $s \to \pm\infty$ are called the free extremity and contact extremity, respectively.

As matching conditions at the free extremity, we assume $n_b \to 0$ and $\partial_s Y \to -sin(\theta_{fx})$, where θ_{fx} is a quantity called the contact angle. Likewise, the membrane tension must satisfy $T(s) \to T_{fx}$, where T_{fx} is the tension on the macroscopic scale outside the contact region.

As a result of the adhesive bonds the membrane is mechanically clamped to the substrate deep within the contact region. Thus we assume that in the limit $s \to +\infty$, n_b and $Y(s,t)$ and $T(s)$ asymptotically approach constant values n_{bcx} and Y_{cx} and T_{cx}, respectively. Remember that, although quantities like θ_{fx} and n_{bcx} are constants when viewed on the fast or microscopic scale of the boundary layer, these quantities can be variables on the macroscopic or global scale.

An important feature of the boundary layer of Figure 1 is that it moves as the edge of the contact zone moves. The normal velocity of this motion at a particular point on the circumference of the contact zone is called the peeling velocity, V_{pl}. We follow the convention that $V_{pl} > 0$ if the motion diminishes the contact region. Subsequently, we will focus our efforts on explicating the theory of the peeling velocity. For the moment however we should only note that V_{pl} is not directly known nor can it simply be specified arbitrarily. In general, it is necessary to obtain V_{pl} implicitly by means of a consistency argument.

3. Thermodynamic Analysis

Preliminary to a more complete analysis of peeling kinetics, it is instructive to consider the static energetics of our theoretical model. The basic framework we employ for such energetic analysis of biological adhesion is the method of Bell, Dembo, and Bongrand (Bell et al., 1984; Dembo and Bell, 1987). Evans was the first to consider the BDB method as applied to the analysis of a peeling membrane, and he obtained some of the results we describe below (see the appendix of Evans, 1985).

In the BDB formalism the membrane plus substrate plus surrounding solvent are viewed as a thermodynamic system closed to fluxes of material substances. The pressure, volume, and temperature of the system are also fixed. Various additional constraints can be considered as interest warrants. In the present case there is an upper constraint on total surface area of contact (i.e., A_{mem}), and in addition the density of adhesion molecules on the membrane and of sites on the substrate is fixed.

The free energy of the closed system is written as a formal sum of various independent contributions:

$$G_{tot} = G_0 + G_m + G_s + G_r + G_b \ . \tag{1}$$

In this equation G_0 is an arbitrary constant containing the free energy of all the components of the system that are unaffected by the occurrence or extent of contact. For example, in the present calculation we will include the free energy of the membrane glycocalyx as being part of G_0. The variable components of the free energy are G_m: the free energy due to surface tension of the membrane, G_s and G_r: the free energy due to unbonded sites and receptors, and finally G_b: the free energy due to adhesive bonds. To avoid confusion for those comparing our current discussion with the treatment in the appendix of Evans (1985), we should mention that, in the latter, G_m was regarded as part of G_0.

Theory of G_m: The surface tension of a membrane is analogous to that of the negative of its free energy per unit area (i.e., the relation between surface tension and area negative pressure with volume). To compute G_m we thus need simply note that at equilibrium the surface tension must be constant except in the thin boundary layer separating the contact region from the remainder of the membrane. Integrating the tension over the total membrane area gives:

$$G_m = -(A_{mem} - A)T_{fx} - AT_{cx} \ . \tag{2}$$

Note that we neglect the energy due to the surface tension of membrane inside the boundary layer because this layer is very thin.

In general it is not possible to directly measure the tension inside a contact zone between two cells or between a cell and a substrate. Nevertheless, it is very easy to deduce the value of T_{cx} from an experimental observation of the contact angle. To derive the relation between T_{cx} and θ_{fx} we need only consider the balance of force parallel to the substrate at the edge of contact. Since, by assumption, bonds orient perpendicular to the substrate, a simple free body diagram of the contact edge suffices to yield a form of Young's equation:

$$T_{cx} = \lim_{s \to +\infty} [T(s)\partial_s X(s)] = - \lim_{s \to -\infty} [T(s)\partial_s X(s)] = -T_{fx} \cos(\theta_{fx}) \ . \tag{3}$$

The validity of Young's equation is well established for contacting fluids and also for the cohesion of other simple materials like lipid vesicles and mica sheets. We feel that it is therefore reasonable to assume that Young's relation also applies to cell adhesion, but we should caution that to our knowledge this has never been checked. In sum, Young's equation is the result of a nontrivial assumption in our model (the assumption that bonds apply stress normal to the plane of the substrate). This assumption is difficult to test, and it is not guaranteed by thermodynamic law.

Theory of G_r and G_s: When surface concentrations are well below the close-packing limit, the expression for chemical potential of unbounded receptors at an arbitrary point on the membrane is:

$$(4)se \qquad \mu_r(n_r) = \mu_r^0 + \beta^{-1} \ln(n_r) \ ,$$

where μ_r^0 is the chemical potential of receptors at unit concentration (one molecule per cm^2) and β is the inverse of the thermal energy.

Equation (4) applies point by point at all locations both inside and outside the contact area. However, if the boundary layer is thin, then for purposes of computing the integrated energy, the receptor density can be regarded as a step function: $n_r = n_{rtot}$ outside the contact area and $n_r = n_{rtot} - n_{bcx}$ inside the contact area. Applying these considerations we can multiply Eq. (4) by n_r and integrate over the membrane to obtain the total contribution of the free receptors to the free energy:

$$(5a) \qquad \begin{aligned} G_r \approx\ & \mu_r^0(A_{mem} - A)n_{rtot} + \beta^{-1}(A_{mem} - A)n_{rtot}\ln(n_{rtot}) \\ & + \mu_r^0 A(n_{rtot} - n_{bcx}) + \beta^{-1}A(n_{rtot} - n_{b,cx})\ln(n_{rtot} - n_{bcx}) \ . \end{aligned}$$

Inside the contact region the membrane and the substrate are parallel since $Y = Y_{cx}$ is constant. Thus the projection of the contact area onto the substrate is area-preserving except in the transition layer. Consequently, n_s, viewed as a function of position on the substrate, can be approximated by a step function of exactly the same form as n_r, viewed as a function of position on the membrane. In other words, the number of bonds is negligible outside the contact region and $n_s = n_{stot}$; inside the contact region, $n_s = n_{stot} - n_{bcx}$. We conclude that for purposes of an energy calculation the presence of the receptors and sites on distinct manifolds is not important and that in consequence an expression analogous to (5a) can be written for the free energy of sites:

$$(5b) \qquad \begin{aligned} G_s \approx\ & \mu_s^0(A_{sub} - A)n_{stot} + \beta^{-1}(A_{sub} - A)n_{stot}\ln(n_{stot}) \\ & + \mu_s^0 A(n_{stot} - n_{bcx}) + \beta^{-1}A(n_{stot} - n_{bcx})\ln(n_{stot} - n_{bcx}) \ . \end{aligned}$$

Lastly, since we are assuming that $n_{stot} \gg n_{rtot} > n_b$, (5b) can be simplified by casting out all but constant terms and terms linear in n_{bcx}:

$$G_s \approx (A_{sub}n_{stot} - An_{bcx})(\mu_s^0 + \beta^{-1}\ln(n_{stot})) - \beta^{-1}An_{bcx} \ .$$

Theory of G_b: In the case of the adhesive bonds the chemical potential at the unit concentration is a variable because bonded molecules are stretched or compressed depending on the size of the gap separating the adherent surfaces. If we model the internal energy of a bond in the fashion of a Hookean spring, then the resulting expression for standard state chemical potential is:

$$\mu_b^0(Y) = \mu_b^0 + 0.5\kappa(Y - L)^2 \quad . \tag{6a}$$

Here, μ_b^0 is the chemical potential of unstressed bonds at unit surface concentration, κ is the spring constant of a bond, L is the rest length of a bond, and Y is the actual length. Supplementing (6a) with the usual term for mixing entropy of a dilute system, we obtain the chemical potential of bonds at arbitrary stress and arbitrary concentration as:

$$\mu_b(n_b, Y) = \mu_b^0 + 0.5\kappa(Y - L)^2 + \beta^{-1}\ln(n_b) \quad . \tag{6b}$$

Multiplying the chemical potential by the bond density and doing the appropriate integral directly, yields the total free energy due to bonds as:

$$G_b = An_{bcx}\left(\mu_b^0 + 0.5\kappa(Y_{cx} - L)^2 + \beta^{-1}\ln(n_{bcx})\right) \quad . \tag{7}$$

As in previous integration steps, we have utilized the fact that $n_b = 0$ outside the contact region and $Y = Y_{cx}$ inside the contact region. We have neglected the contributions from the transition boundary layer.

Minimizing the free energy: At equilibrium, the total free energy must be minimized with respect to the available degrees of freedom: n_{bcx}, Y_{cx}, and A. Differentiating G_{tot} with respect to Y_{cx} and setting the result equal to zero yields:

$$\partial_{Y_{cx}} G_{tot} = 0 = An_{bcx}\kappa(Y_{cx} - L) \quad . \tag{8a}$$

We conclude that at equilibrium the gap between the cell and the substrate inside the contact zone is equal to the unstressed bond length.

Differentiating G_{tot} with respect to n_{bcx} and setting the result equal to zero, we obtain an equation for the bond density inside the contact zone as:

$$n_{bcx} = (n_{rtot} - n_{bcx})n_{stot}K_l \exp\left[-0.5\beta\kappa(Y_{cx} - L)^2\right] \quad . \tag{9a}$$

In this equation

$$K_l \equiv \exp\left[1 - \beta(\mu_b^0 - \mu_r^0 - \mu_s^0)\right] \tag{9b}$$

is the equilibrium constant for formation of an unstressed bond. The units of K_l are the inverse of the units in which surface concentration is measured: usually $cm^2/molecule$. The complete equilibrium constant for formation of a bond, regardless of the state of stress, is seen to be the product of K_l and a Boltzmann correction containing the strain energy:

$$K(Y) \equiv K_l \exp\left[-0.5\beta\kappa(Y-L)^2\right] \quad . \tag{9c}$$

Using the fact that $Y_{cx} = L$, we can now easily solve Eq. (9a) to obtain the equilibrium bond density in the contact region as:

$$n_{bcx} = \frac{K_l n_{rtot} n_{stot}}{(1 + K_l n_{stot})} \quad . \tag{10}$$

Finally, differentiating G_{tot} with respect to A and using Eq. (10) we obtain:

$$\begin{aligned} \partial_A G_{tot} &= (1+\cos(\theta_{fx}))T_{fx} + \beta^{-1} n_{rtot} \ln(1 - n_{bcx}/n_{rtot}) \\ &= (1+\cos(\theta_{fx}))T_{fx} - \beta^{-1} n_{rtot} \ln(1 + n_{stot} K_l) \quad . \end{aligned} \tag{11}$$

Notice that the right side of this equation has no explicit dependence on the contact area. This means that in general the equation $\partial_A G_{tot} = 0$ can be satisfied for any value of A but only provided the tension on the free extremity of the membrane is exactly equal to a certain critical value:

$$T_{crit} = \frac{\beta^{-1} n_{rtot} \ln(1 + n_{stot} K_l)}{(1+\cos(\theta_{fx}))} \quad . \tag{12}$$

Some experimental studies confirming certain aspects of Eq. (12) have recently appeared (Kuo and Lauffenburger, 1993).

The preceding excursion into the realm of membrane equilibria, culminating with Eq. (12) for the critical tension, is the minimum needed for our subsequent discussion of peeling and attachment. Needless to say, the subject of critical tensions and thermal equilibrium is of interest in its own right. The reader may therefore find it an interesting exercise to generalize our expression for the critical tension to include the case $n_{stot} \leq n_{rtot}$. Also of interest are cases where the receptors are mobile, the sites are mobile, or where both receptors and sites are diffusible. Finally, all these cases can be compounded with the presence of a finite repulsive potential of the glycocalyx. The last avenue of generalization reveals new phenomena because it implies that $Y_{cx} > L$.

From Eq. (11) it can be seen that if $T_{fx} < T_{crit}$ then G_{tot} is always a decreasing function of A. Theoretically this means no state of thermal equilibrium exists and that the membrane will continue attaching to the substrate until the constraint $A \leq A_{mem}$ is violated. Conversely, if $T_{fx} > T_{crit}$ then $\partial_A G_{tot} > 0$ for all A. Thus the membrane will continuously peel until it is completely detached.

We conclude that equilibrium thermodynamics provides a satisfactory analysis of our model only under the very special condition $T_{fx} = T_{crit}$. In order to give some reasonable account of what happens when $T_{fx} \neq T_{crit}$, the central question is the physics of the frictional dissipation or entropy production associated with changes in adhesion. In addition to our own work (cf. next section) various other methods of thinking about this problem have been tried. For example, a very promising approach involves extension of the equilibrium BDB method into the regime of nonequilibrium thermodynamics (Zhu, 1991). Equally interesting is the technique of direct stochastic simulation (Hammer and Lauffenburger, 1987).

4. Chemical Kinematics in the Contact Frame

Let us return to the moving boundary layer of Figure 1 and, in the associated coordinates, consider a position on the membrane manifold corresponding to arc length coordinate s and the projection of this point on the substrate, $X(s)$. The formation and breakage of cell-to-surface bonds between these points can be regarded as a stochastic chemical rate process of the type

$$n_r(s) + n_s(X(s)) \underset{k_r(Y)}{\overset{k_f(Y)}{\rightleftharpoons}} n_b(s) \quad , \tag{13}$$

where the rate constants $k_f(Y)$ and $k_r(Y)$ are assumed to be given by some kind of constitutive functions of the membrane-to-substrate separation distance (specified subsequently).

Because receptors are fixed in the plane of the membrane and sites are fixed in the plane of the substrate, the above reaction is subject to conservation of receptors:

$$n_r(s) = n_{rtot} - n_b(s) \quad , \tag{14a}$$

and conservation of sites:

$$n_s(X(s)) = n_{stot} - n_b(s)[\partial_s X(s)]^{-1} \quad . \tag{14b}$$

Note that to be completely rigorous in the treatment of the latter constraint, we have to include a correction for the area contraction associated with the projective mapping of the membrane manifold onto the substrate. This is a cute detail, but it leads to a mathematical nonlinearity that obscures more fundamental physics. Therefore, in subsequent developments we will restrict consideration to the approximate form of Eq. (14b) obtained in the obvious way by assuming $n_{stot} \gg n_{rtot}$. After this simplification we can use Eqs. (14a) and (14b) to write the complete continuity equation for bond density in the contact frame as:

$$\partial_t n_b \approx V_{pl} \partial_s n_b + k_f(Y) n_{stot} (n_{rtot} - n_b) - k_r(Y) n_b \quad . \tag{15a}$$

When interpreting the last equation, it is essential to remember that the peeling velocity does not correspond to a real convection of bonds but is simply a virtual convection due to the relative motion of the contact frame with respect to the laboratory. This motion of the contact frame is determined by the purely logical requirement that the bond density be equal to the threshold density at the origin of arc length. Given that n_b satisfies this requirement at $t = 0$, consistency at all subsequent times will be maintained if and only if $\partial_t n_b = 0$ at $s = 0$. In view of Eq. (15a), the latter can be restated in the form of an explicit formula for the peeling velocity:

$$(15b) \qquad V_{pl}(t) = \left[\frac{k_f(Y) n_{stot} \left(n_{rtot} - n_b \right) - k_r(Y)\, n_b}{\partial_s n_b} \right]_{s=0} .$$

5. Catch Bonds, Slip Bonds, and Ideal Bonds

Equation (15b) is incomplete because the rate constants (or more accurately the rate coefficients) of the bonding reaction are arbitrary functions of the vertical separation between the membrane and the surface. To move beyond such ambiguity, it is first of all necessary to restrict the choice of $k_f(Y)$ and $k_r(Y)$ by requiring that these quantities maintain consistency with what we have previously derived concerning the equilibrium constant of bonding [i.e., Eq. (9c)]. In other words, for any value of Y we must require that the ratio of k_f to k_r satisfy:

$$(16) \qquad \frac{k_f(Y)}{k_r(Y)} \equiv K(Y) = K_l \exp(-0.5\beta\kappa(Y - L)^2) \ .$$

Thus, $k_f(Y)$ and $k_r(Y)$ are not independent but can be determined one from the other in terms of thermodynamic constants K_l, κ, and L.

To complete the development of the constitutive behavior of k_r and k_f we follow a line of argument made familiar by the well known Arrhenius theory of chemical reaction rates. In this theory a dissociation reaction is described in terms of motion along a pencil of optimal or near-optimal pathways traversing the energy surface in configuration space and connecting the bonded reactant and the dissociated products. The rate for the overall dissociation is naturally limited by the time spent on the most arduous portion of this journey, i.e., near to the saddle point of the energy surface. This saddle point, also called the "transition state," is defined as the highest point lying on the optimal pathway connecting the reactant and products (i.e., the worst point on the optimal path).

The inverse of the characteristic decay time of the transition state is called the thermal frequency, k_0. By a simple quasi-steady-state argument, it is easy to show that the overall dissociation rate of an adhesive bond can be approximated as the product of the thermal frequency and a Boltzmann factor that gives the relative likelihood of the transition state and the bonded state at equilibrium:

$$k_r \approx 0.5 k_0 \exp(-\beta E_a)\ . \tag{17}$$

The activation energy appearing in the Boltzmann term is defined as the difference in chemical potential between the transition state and the bound state. Because the transition state is a saddle point, its decay requires a symmetry breaking that can depend only on the effective inertial mass of the transition state, the frictional coupling with the solvent, and the correlation time of the random thermal forces. The essential conclusion for purposes of our analysis is that the thermal frequency is independent of bond strain. Thus, if we wish to know how k_r depends on Y, we must look to the influence of strain on E_a.

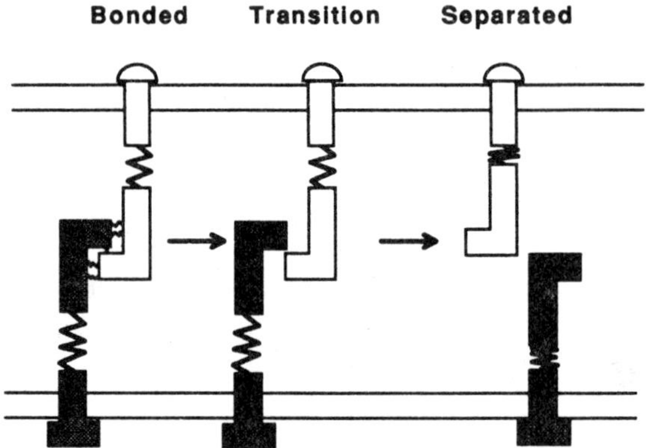

Figure 2. The bonded state, the transition state and the separated state for disruption of an adhesive crosslink.

The transition state for adhesive bonding has a rather well defined structure; it occurs after the entropic and enthalpic pain involved in bringing the two opposite binding domains into close proximity and before the gain from the stabilizing non-covalent interactions. We therefore conclude that the transition state of an adhesive bond is characterized by a definite stiffness and length just like the final bonded state.

The necessary strain analysis of the activation energy is somewhat simplified because biological adhesion is generally mediated by large macromolecules. These macromolecules can be decomposed conceptually into two parts or domains: a relatively small "binding domain" and a large "linking domain." The binding domains of the receptors and the sites have typical dimension of $\approx 10\mathring{A}$ and contain the moeities involved in the noncovalent interactions that energetically stabilize the bonded state. The linking domains have dimension of $\approx 100\mathring{A}$ and, by virtue of this sheer bulk, they control the overall mechanical stiffness and size of the receptors and sites.

Regardless of how one chooses to visualize the detailed arrangement of the binding and linking domains, the transition state for adhesive bonding has a rather well defined structure; it occurs after the entropic and enthalpic pain involved in bringing the two opposite binding domains into close proximity and before the gain from the stabilizing non-covalent interactions (see Figure 2). We therefore conclude that the transition state of an adhesive bond has a certain stiffness and length just like the final bonded state. Thus in analogy with Eq. (6a) we can write a Taylor's series for the free energy of the transition state

$$\mu_{ts}^0(Y) = \mu_{ts}^0(L_{ts}) + 0.5\kappa_{ts}(Y - L_{ts})^2 + \dots \quad , \tag{18}$$

where L_{ts} is the unstressed length of the transition state and κ_{ts} is the spring constant of the transition state.

Combining Eqs. (6a) and (18), the activation energy for rupture of an adhesive bond is given by:

$$\begin{aligned} E_a = \mu_{ts}^0(Y) - \mu_b^0(Y) =& (\mu_{ts}^0(L_{ts}) - \mu_b^0(L)) \\ &+ 0.5\kappa_{ts}(Y - L_{ts})^2 - 0.5\kappa(Y - L)^2 + \dots \quad . \end{aligned}$$

Because the bonding domains of biological adhesion molecules are much smaller than the linking domains and because the bonding domains are already in close proximity in the transition state, we surmise that the stiffness and length of the transition state must be very close to the corresponding properties of the final bonded state (i.e., $L_{ts} \sim L$ and $\kappa_{ts} \sim \kappa$). This mundane observation is very helpful because it leads us to expand the activation energy for bond rupture as a triple Taylor's series:

(19)

$$\begin{aligned} E_a \equiv& \mu_{ts}^0(Y) - \mu_b^0(Y) \approx \mu_{ts}^0(L_{ts}) - \mu_b^0(L) - \kappa(L_{ts} - L)(Y - L) \\ &+ O((\kappa_{ts} - \kappa)(Y - L)^2) + O((L_{ts} - L)(\kappa_{ts} - \kappa)(Y - L)) + \dots \quad . \end{aligned}$$

Substituting this into Eqs. (16) and (18) and casting out higher order terms, we arrive at the final result:

$$k_r(Y) \approx k_{rl}\exp(\Delta_l\sqrt{\kappa\beta}\,(Y - L)) \quad , \tag{20a}$$

and

$$k_f(Y) \approx k_{fl}\exp(\Delta_l\sqrt{\kappa\beta}\,(Y - L) - 0.5\kappa\beta(Y - L)^2) \quad , \tag{20b}$$

where $\Delta_l \equiv \sqrt{\beta\kappa}\,(L_{ts} - L)$ and where k_{fl} and k_{rl} are the forward and reverse rate constants for formation of unstressed bonds. Writing the tensile force per bond as $f_b = \kappa(Y - L)$, it is seen that these constitutive laws can be recast in such a way as to express the forward and reverse rate constants in terms of bond stress as opposed to bond strain.

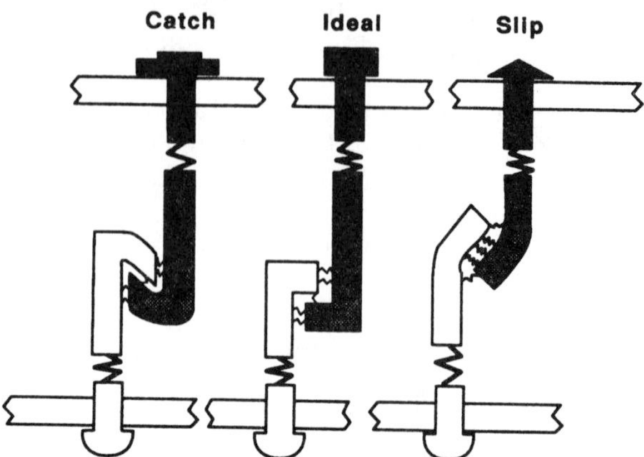

Figure 3. Classification of adhesion molecules as either catch bonds, slip bonds, or ideal bonds.

According to the theory outlined in the text, a bond is a "catch" bond if its transition state has a shorter rest length than its bonded state (i.e., if $\Delta_l < 0$). According to Eq. (20a) this implies that the rate of disruption after stretching will actually be slower than the rate before stretching.

Slip bonds occur if the transition state is longer than the bonded state (i.e., if $\Delta_l > 0$). According to Eq. (20a) this condition means that the application of tensile stress will accelerate the rate of bond rupture.

Ideal bonds represent the intermediate case: $\Delta_l = 0$. In this case the rate of rupture becomes independent of strain or stress.

Implicit in the preceding theory of bond rupture is a classification of the possible modalities by which biological adhesion molecules can respond to mechanical stress (see Figure 3). To make this explicit it is helpful to first define what is meant by an "ideal" adhesive bond and then to consider departures from the ideal case (Dembo et al., 1988). Accordingly, we say a biological adhesion molecule is "ideal" if the mechanical coefficients (rest length and spring constant) of its transition state are all exactly equal to the corresponding properties of its bonded state (in the current example this means that perturbation parameter Δ_l is exactly equal to zero). In this ideal limit it can be seen from Eq. (20a) that the reverse rate constant is completely independent of the applied stress; whereas the forward rate constant still remains strongly dependent on Y.

Extending the above argument, the degree and manner with which the reverse rate constant of a real adhesive bond changes in response to mechanical stress provides a direct experimental measure of its departure from ideality. For most biological adhesion molecules, we expect that these departures will be small since we have argued that Δ_l is small. If this is true then we must predict that the rate constant for rupture of an adhesive bond will generally be much less sensitive to mechanical stress than the rate constant for formation of the bond.

When the transition state is shorter than the bound state (i.e., if $\Delta_l < 0$), then Eq. (20a) implies that the rate of bond disruption after stretching will actually be slower than the rate for disruption before stretching. At first glance this might seem rather implausible, but in actuality it is easy to construct molecular models of cell-to-substratum bonds that will give this behavior. The basic mechanism of such models is similar to that of a hook or a barb. It can also be viewed as analogous to the well known child's toy called a "finger prison." In other words, the structure redirects the stress created by stretching the bond so as to lock the bonding groups more tightly together. To evoke this image in our subsequent discussions we will refer to bonds with $\Delta_l < 0$ as "catch-bonds." According to Eq. (20a) if $\Delta_l > 0$, then the application of tensile stress will accelerate the rate of bond rupture. Bonds that display this kind of intuitively natural behavior are classified as "slip" bonds.

6. Mechanics in the Contact Frame

Because we regard bonds as mechanically equivalent to Hookean springs oriented perpendicular to the substrate, the normal and tangential tractions they exert on the membrane can be written as:

$$\sigma_{nor} = +n_b\kappa(Y - L)\partial_s X \tag{21a}$$

and

$$\sigma_{tan} = -n_b\kappa(Y - L)\partial_s Y \ , \tag{21b}$$

respectively. The normal and tangential force balances on the membrane in the transition zone result in the nonlinear system (Evans and Skalak, 1980; Evans, 1985)

$$\partial_s(T + 0.5B_m C^2) = -\sigma_{tan} \tag{22a}$$

and

$$B_m\partial_s^2 C - CT = -\sigma_{nor} \ , \tag{22b}$$

where B_m is the modulus of bending, C is the curvature, and T is the tension.

In addition to the above, we know by definition of the curvature that:

$$C = (\partial_s X)(\partial_s^2 Y) - (\partial_s Y)(\partial_s^2 X) \ , \tag{22c}$$

and by definition of the arc length that:

$$(\partial_s X)^2 + (\partial_s Y)^2 = 1 \quad . \tag{22d}$$

Equations (21a) and (21b) for the bond stresses and (22a)-(22d) for mechanical equilibrium are, of course, supplemented by the continuity equation of (15a), by the kinematic condition for the peeling velocity (15b), by the constitutive laws for the chemical rate coefficients (20a) and (20b), by initial conditions on n_b, and by boundary conditions at the free and clamped extremities. The result is a well posed mathematical system for the primitive variables $V_{pl}(t)$, $n_b(s,t)$, $T(s,t)$, $X(s,t)$, and $Y(s,t)$ in the boundary layer separating adherent membrane from nonadherent membrane. We now consider the results obtained from study of some solutions of these equations.

7. The Existence and Stability of States of Steady Peeling

There is no reason to expect *a priori* that our model admits of stable solutions corresponding to states of steady peeling or attachment. As examples of other possibilities, membrane in the transition zone could buckle, leading to infinite peeling velocities after a short transient. Alternatively, oscillatory cycles of rapid and slow peeling could continue indefinitely. Finally, the peeling velocity could continuously increase or decrease remaining finite except in the limit $t \to \infty$, but without ever reaching an equilibrium.

Despite these manifold dire possibilities, we have previously shown by direct numerical computation that, at least in the case of ideal bonds and slip bonds, a state of steady peeling or attachment exists and is stable for all tensions. An illustrative example is shown in Figure 4. Computations corresponding to peeling ($T_{fx} = 2T_{crit}$), equilibrium ($T_{fx} = T_{crit}$), and attachment ($T_{fx} = 0.5T_{crit}$) are given.

The initial transient phase during which the boundary of the contact zone undergoes considerable acceleration can be seen. To show how the results depend on the choice of contact threshold, each computation was carried out for $\delta_b = 10^{-4}n_{bcx}$ (open symbols) and for $\delta_b = 10^{-6}n_{bcx}$ (closed symbols). During the transient phase, V_{pl} depends slightly on the choice of contact threshold. However in all cases, a steady state is eventually reached wherein the peeling velocity becomes independent of the choice of contact threshold.

In the case $T_{fx} \leq T_{crit}$, we find that the approach to the final steady state is always direct, rapid, and monotone. Such simple kinetics are also found for peeling when the reactive compliance of the bonds is large (i.e., $\Delta_l > 1$) or when equilibrium binding is weak ($n_{stot}K_l << 1$). In the general case, however, the kinetics of detachment consist of a lag phase, followed by abrupt acceleration (which can even overshoot the stable peeling rate), and finally an approach to the stable peeling velocity through a sequence of damped oscillations. This implies that in the case of peeling, the eigenvalues of the linear perturbation of steady peeling are complex.

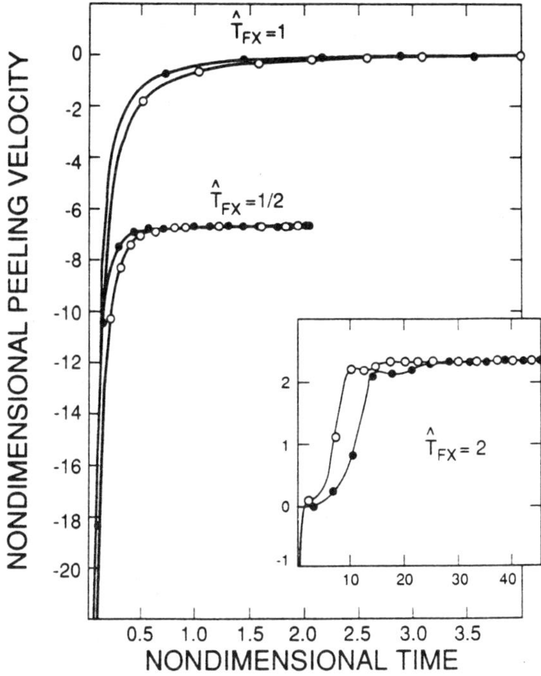

Figure 4. Computations of the time dependence of the peeling velocity in the case of ideal bonds.

Nondimensional time and nondimensional peeling velocity are defined as $[k_{rl}t]$ and $[\sqrt{\beta\kappa}V_{pl}/k_{rl}]$, respectively. In all computations the initial distribution of bonds was taken to be:

$$
\begin{aligned}
n_b &= 0 \quad &if\ s < 0 \ ,\\
n_b &= \delta_b \quad &if\ s = 0 \ ,\\
n_b &= n_{bcx} \quad &if\ s > 0 \ .
\end{aligned}
$$

Three pairs of curves are shown corresponding to $\hat{T}_{fx} = [T_{fx}/T_{crit}] =$ 0.5, 1.0, and 2.0. Within each pair, one curve corresponds to contact threshold $\delta_b = n_{bcx}10^{-4}$ (open symbols) and the other to $\delta_b = n_{bcx}10^{-6}$ (closed symbols). In all computations $[n_{rtot}/(B_m\kappa\beta^2)] = 1.68 \times 10^{-4}$ $[\theta_{fx}] = \pi/2$, and $[n_{stot}K_l] = 1$.

The results demonstrate the existence of a state of steady peeling that is approached asymptotically at long times for any choice of the tension. Once this state is attained, the peeling velocity does not depend on the choice of contact threshold.

The results of Figure 4 are generally descriptive of ideal bonds and slip bonds; however, the case of catch bonds reveals new qualitative behavior. Specifically, we find numerically that if bonds are of catch-type (i.e., $\Delta_l < 0$) and if the tension is greater than the critical tension, then no matter how other parameters are chosen, the membrane will start to peel but will eventually grind to a halt. The exact time required before the "catch" state is reached depends on the tension and on exactly how negative the reactive compliance of the catch bonds happens to be. This is the only qualitative difference between catch bonds and other bond classifications. In attachment or equilibration modes (i.e., $T_{fx} \leq T_{crit}$), catch bonds are characterized by a monotonic initial transient followed by a steady attachment of the membrane to the substrate that continues indefinitely.

In summary, the equations governing the formation of catch-bonds seem to be (and in fact are) fully reversible in a thermodynamic sense. Nevertheless, although the membrane can anneal to a surface, it is impossible to peel it off at a finite velocity without doing irreversible damage. To get an intuitive feeling for the predicted properties of catch-bond mediated biological adhesion, it is helpful to think in terms of the well known synthetic material called "velcro."

The aforementioned results settle the question of the existence of states of steady peeling. Such states exist for slip bonds and ideal bonds; such states do not exist for catch bonds. In those cases where states of steady peeling exist, it then becomes important to know the actual value of the steady state peeling velocity.

8. Steady Peeling of Ideal Bonds

Figure 5 presents a typical numerical solution showing the tension, deflection, and bond density in the contact boundary layer during peeling. As can be seen from this figure, the deflection of the membrane as it approaches the substrate in the adhesion boundary layer is approximately exponential:

$$Y(s) \approx L + D_0 \exp[-s/\lambda] \quad , \tag{23a}$$

where D_0 is the deflection at the edge of the contact zone and λ is a characteristic dimension of the transition zone (to be specified subsequently).

We now define a nondimensional displacement:

$$\zeta(s) = \sqrt{\beta\kappa}\ [Y(s + V_{pl}/k_{rl}) - L] = \zeta_0 e^{-\epsilon}\, e^{-s/\lambda} \quad , \tag{23b}$$

where $\zeta_0 \equiv D_0\sqrt{\beta\kappa}$ and $\epsilon \equiv V_{pl}/(k_{rl}\lambda)$ is the nondimensional peeling velocity.

Setting time derivatives equal to zero, setting $k_r = k_{rl}$, and introducing ζ as a new spatial variable, we find that Eq. (15a) (continuity equation) can be written as:

$$-\epsilon\,\zeta\partial_\zeta \tilde{n}_b = -\tilde{K}(\zeta\, e^{\epsilon}) + (1 + \tilde{K}(\zeta\, e^{\epsilon})\,)\tilde{n}_b \quad , \tag{24}$$

where $\tilde{n}_b \equiv n_b/n_{rtot}$ and $\tilde{K}(\zeta) \equiv n_{stot}K_l \exp[-0.5\zeta^2]$.

We now obtain a power series solution in the usual way.

$$\tilde{n}_b(Y,\epsilon) = \frac{\tilde{K}(\zeta)}{1+\tilde{K}(\zeta)}\left[1 + \frac{2\epsilon\tilde{K}(\zeta)\ln[\tilde{K}(\zeta)/\tilde{K}(0)]}{(1+\tilde{K}(\zeta))^2} + O(\epsilon^2)\right] . \tag{25}$$

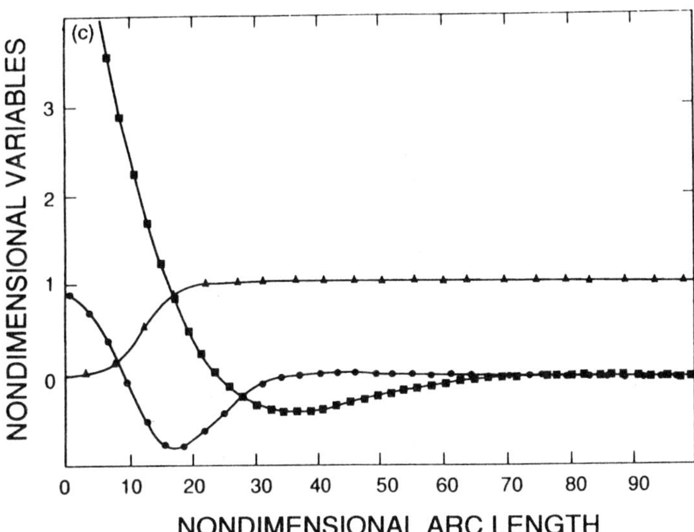

Figure 5. Spatial variation of deflection, tension, and cross bridge density in the contact boundary layer during steady peeling.

Arc length is given in nondimensional form as $[s\sqrt{\beta\kappa}]$. Deflection is shown as $[\sqrt{\beta\kappa}(Y(s) - L)]$ (squares), tension as $[T(s)/T_{crit}]$ (circles), and bond density as $[n_b/n_{bcx}]$ (triangles). For this calculation the nondimensional tension at the free extremity was $[T_{fx}/T_{crit}] = 2$, and the nondimensional time was allowed to approach infinity. Other nondimensional parameters are the same as in Figure 4.

For the subsequent analysis, the essential observation about this series is that it converges uniformly in ζ. In fact, the need to obtain a series with this property is the motivation for the artifice of including the peeling velocity in the definition of ζ. The reader can verify that if this factor is omitted, then the percent error of a power series solution has a singularity when $\zeta \to \infty$ (i.e., at the free extremity).

We now use Eq. (25) to obtain an approximate expression for the tangential stress [Eq. (21a)].

$$\sigma_{tan} = -\left(\frac{n_{rtot}e^{2\epsilon}}{\beta}\right)\tilde{n}_b\zeta\partial_s\zeta \approx \left(\frac{n_{rtot}e^{2\epsilon}\tilde{n}_b}{\beta\tilde{K}}\right)\partial_s\tilde{K} \quad . \tag{26}$$

Substituting for σ_{tan} in the equation of tangential force balance [Eq. (22a)] and integrating with respect to s, yields:

(27)

$$T_{cx} - T_{fx} = -\int_{-\infty}^{+\infty} \sigma_{tan} ds$$

$$\approx -\left(\frac{n_{rtot}e^{2\epsilon}}{\beta}\right)\left[\ln(1+\tilde{K}(0)\,) - \frac{\epsilon 0.25\ln(2)\tilde{K}(0)^2}{(1+\tilde{K}(0)\,)^3} + O(\epsilon^2)\right] .$$

Recalling Young's relation [Eq. (3)] and the definition of the critical tension [Eq. (12)], we now invert Eq. (27) to obtain:

$$V_{pl} \approx 0.5k_{rl}\lambda\ln(\frac{T_{fx}}{T_{crit}})\;(1+O(\epsilon)\,) \quad . \tag{28}$$

Equation (28) determines the peeling velocity except for the fact that the value of λ is unknown. To correct this deficiency we must recall that by its definition λ is the characteristic wavelength that governs the final approach of the membrane to the substrate at the inner edge of the transition zone [see Eq. (23)]. An estimate of λ in the current model can be obtained by considering Eq. (22b) (normal force balance) in the limit $s \to +\infty$. This limit facilitates a linear analysis because $n_b \approx n_{bcx} =$ *constant* and $T \approx T_{cx} =$ *constant*. Furthermore, from Eq. (22d) (definition of arc length) and Eq. (22c) (definition of the curvature), we also estimate that

$$\lim_{s\to+\infty} \partial_s X(s) = \sqrt{1-(\partial_s Y(s)\,)^2} \approx 1 - 0.5\lambda^{-2}(Y(s)-L)^2 \approx 1 \quad ,$$

$$\lim_{s\to+\infty} C \approx \partial_s^2 Y(s) + O((Y(s)-L)^2) \approx \lambda^{-2}(Y(s)-L) \quad ,$$

and

$$\lim_{s\to+\infty} \partial_s^2 C \approx \partial_s^4 Y(s) + O((Y(s)-L)^2) \approx \lambda^{-4}(Y(s)-L) \quad .$$

Substituting into Eq. (22b) and canceling common factors gives the characteristic polynomial:

$$0 = B_m\lambda^{-4} - \lambda^{-2}T_{cx} + n_{bcx}\kappa \quad . \tag{29}$$

For purposes of our analysis we are only interested in the decay constant of the most slowly changing mode for positive values of the membrane tension. If $0 < T_{cx} < +2\sqrt{B_m n_{bcx}\kappa}$, then the roots of (29) are complex corresponding to damped spatial oscillations of the membrane as it approaches the substrate. The decay constant for this class of solutions is

$$\Re(\lambda^{-1}) \approx \left[(\frac{T_{cx}}{4B_m}) + \sqrt{\frac{n_{bcx}\kappa}{4B_m}}\,\right]^{\frac{1}{2}} .$$

Finally, if $+2\sqrt{B_m n_{bcx}\kappa} < T_{cx}$, then the roots of (26) are all real. The most slowly decaying mode gives

$$\Re(\lambda^{-1}) \approx \left[(\frac{T_{cx}}{2B_m}) - \sqrt{(\frac{T_{cx}}{2B_m})^2 - (\frac{\kappa n_{bcx}}{B_m})}\,\right]^{\frac{1}{2}} .$$

Combining the special cases, we derive a continuous approximation that is accurate to within a few percent for $0 < T_{cx} < \infty$:

$$\lambda^{-1} \approx \sqrt{-(\frac{T_{cx}}{B_m}) + \sqrt{(\frac{0.5\kappa n_{bcx}}{B_m}) + (\frac{T_{cx}}{B_m})^2}} \quad . \tag{30}$$

Equation (30) completes the analytic theory of the peeling velocity of ideal bonds begun by Eq. (28). Unfortunately the logic of this theory is only approximate and firm error estimates are not available. In such a situation the most efficient method for checking the validity of Eqs. (28) and (30) is by way of a numerical analysis. The methods required for obtaining numerical solutions are nontrivial, but such details are described elsewhere (Dembo et al., 1988). To briefly summarize the essential conclusions, we find that the results of our computations are represented by the following empirical formula for the peeling velocity:

$$V_{pl} \approx \frac{0.5k_{rl}\ln(T_{fx}/T_{crit})[1 + T_{crit}/T_{fx} + \ldots]}{\pi\sqrt{-(T_{fx}/B_m) + \sqrt{(0.5\kappa n_{bcx}/B_m) + (T_{fx}/B_m)^2}}} \quad . \tag{31}$$

It can be seen that this equation is consistent with Eqs. (28) plus (30) except for three aspects: In Eq. (28), T_{cx} is replaced by T_{fx}, in Eq. (30) the term $(1 + O(\epsilon)$ is replaced by $(1 + Tcrit/Tfx + \ldots)$, and finally the right side of Eq. (31) is multiplied by a factor of $1/\pi$. These discrepancies are undoubtedly the result of limitations of the analytic approximations. Thus while the analytic treatment is useful for gaining physical insight, the use of the numerical result, i.e., Eq. (31), is recommended for comparison with experimental data.

Eq. (31) is accurate to within $\approx$ 10 percent even if $\epsilon \leq 5$. No basic qualitative problems with Eq. (31) could be detected with respect to the dependence of peeling velocity on any parameter: membrane tension, contact angle, equilibrium constant, bending modulus, density of adhesion molecules, or bond stiffness. Finally, the numerical results indicate that for ideal bonds significant errors occur only if $T_{fx}/T_{crit} << 1$ and $n_{stot}K_l >> 1$ (i.e., during rapid annealing of the membrane and the substrate when binding is very strong). Under such circumstances Eq. (31) underestimates the rate of annealing. If one is interested only in peeling of ideal bonds, then it is quite accurate to approximate the numerator by retaining only the logarithmic dependence on T_{fx}/T_{crit}.

9. Steady Peeling of Slip Bonds

As described previously, the degree to which bonds depart from ideal behavior is parameterized by a nondimensional group called the reactive compliance, Δ_l. If the reactive compliance is small, then it is easy to show that this term does not affect the leading behavior of Eq. (28). Numerical studies indicate that the first order correction to the peeling velocity is obtained by multiplying the right of Eq. (31) by a correction factor $[1 + O(\Delta_l T_{fx}/T_{crit})]$. Thus if the bonds are slip bonds and if the peeling tension is made sufficiently high, then this term, linear in the peeling tension, will eventually become important in the numerator of Eq. (31).

We conclude that the general asymptotic behavior of the peeling velocity at large tensions is of the form:

$$V_{pl} \sim T_{fx}^{m} \ln[T_{fx}] \quad ,$$

where the constant $m < 1$ in the case of ideal bonds and $m > 1$ in the case of slip bonds. This last is a relevant point for the analysis of experimental data on the velocity of cell rolling (cf. next section). The observation of a peeling velocity that increases at a sublinear rate at large values of the tension, is diagnostic of bonds that are very close to ideal bonds. Conversely, the observation of a peeling velocity that increases at a rate faster than linear is indicative of slip bonds.

10. Discussion

It is clear from our results that the functional classification of adhesive crosslinks as either catch bonds, slip bonds or ideal bonds is potentially very important to the understanding of biological adhesion at the molecular level. Catch bonds are mechanically irreversible linkages that cannot be "peeled apart" without doing at least some irreversible damage (e.g., uprooting receptors from the membrane). In contrast, once the critical tension is exceeded, slip bonds have very little capacity to resist the kinds of mechanical stresses that might cause peeling. The borderline case is that of an ideal bond: it is possible to mechanically peel macroscopic adherent regions, but at least the bonds offer considerable resistance.

The existing micromanipulation data on cell peeling and attachment have been reviewed recently by Evans (1993). The bulk of the data are consistent with the hypothesis that the underlying biological adhesion molecules form catch bonds. In general, it is observed that very small tensions are sufficient to prevent the growth of established contact regions, and frequently considerable coaxing and impingement is required just to initiate adhesion (i.e., low formation energy). Nevertheless, an established adhesive bond cannot be separated unless very large tensions are applied (high disruption energy).

An additional ubiquitous observation is that the tension required to sustain peeling increases as the extent of peeling progresses. Thus, unlike the situation with synthetic adhesives (Krenceski et al., 1986), there is no point at which a true propagating fracture of the adhesive bond occurs. We find this phenomenology to be very suggestive of catch bonds. As we have described, if a large peeling tension is applied to such bonds, then it simply takes longer for a sufficient number of bonds to simultaneously "catch" and counterbalance the tension. Thus the data demonstrate thermodynamic irreversibility and a failure of fracture propagation for the bonding mediated by a wide range of biological adhesion molecules. We should re-emphasize that, although these data are consistent with the catch bond hypothesis, other interpretations are equally plausible.

As one example of an alternative theory to explain the apparent thermodynamic irreversibility of biological adhesion, consider the cryptic receptor model (Evans et al., 1991). This model relies on the assumption that the membrane in the contact region between two cells has some degree of submicroscopic roughness. It is further assumed that the roughness is sufficient to sterically prevent the adhesion receptors trapped in the valleys between wrinkles from reaching complementary sites. Since peeling is associated with an increase in the surface tension near the edge of the contact region, it is possible that this could decrease the degree of roughness. This mechanism can be visualized as similar to pulling wrinkles out of a bedsheet by applying tension at the edges. The smoothing action of peeling could progressively bring more and more of the adhesion molecules in the remaining contact region of the two surfaces into position for actual binding. Thus the act of peeling causes a feedback that progressively increases the resistance to subsequent peeling.

Given that the simple observation of irreversibility can have many explanations, we have devised what we think is a more definitive test of the catch bond hypothesis. The basic idea of our test rests on the availability of soluble competitive ligands that block the interaction of adhesion receptors and complementary sites. For example, in the case of the selectins certain sugar moeities would constitute such blockers (Springer and Laskey, 1991).

To understand our experiment, consider a cell and a substrate in adhesive equilibrium in the presence of external tension (or some other force) tending to cause peeling. Regardless of one's model, it is clear that at the edge of adhesive contact the bonds are highly strained since, by definition, they must be at the limit of their functioning. If the bonds are slip bonds or ideal bonds, then these highly stressed linkages will be in a rapid equilibrium, constantly forming and breaking. On the other hand, if the bonds are catch bonds, then the reverse rate constant drops to zero when the bonds are under stress [Eq. (20a)]. As a result the bonds at the edge of contact will be in the so-called "catch state," effectively locked in the bonded configuration with very slow off rate.

Competitive blockers act by preventing the formation of new bonds; they have no influence on bonds that already exist. Thus, if one adds a high enough concentration of blocker and if the bonds at the edge of the adhesion region are catch bonds, then we predict that the blocker will be unable to intercalate, and peeling will not be initiated unless very high levels of blocker are employed for very long times. Paradoxically, the greater the peeling tension tending to cause disruption of the catch bonds, the more resistant the bonds should become to the effects of blocker. On the other hand, if the bonds at the edge of contact are slip bonds or ideal bonds, then the blocker should easily be able to intercalate into the adhesive contact, disrupt these bonds, and actuate the peeling process.

Although most studies of biological adhesion are characterized by irreversibility, there is one important exception: the phenomenon of leucocyte rolling (Atherton and Born, 1972; Lawrence and Springer, 1991). Clearly, in the case of rolling, bonds are peeled on some segments of the circumference of the contact region and attached on other segments. If the leucocyte were attached to the substrate by catch bonds, then at least some of these bonds would be irreversibly lost at the trailing edge and rolling could not be sustained indefinitely. Thus it is likely that the bonds mediating leucocyte adhesion to the endothelium are either ideal bonds or slip bonds.

Thus far our analysis has focused on the so-called "inner" part of a classical boundary layer expansion. To compute the rolling velocity, it is necessary to match this with an "outer" solution. For a given contact region, the outer solution consists of a macroscopic description of the various loads, stresses, and deformations of the cell. The solution of this model is used to derive the limiting values of the tension and contact angle at all points on the periphery of the contact region in terms of macroscopic observables such as blood velocity, cell diameter, etc. Equation (31) is then used to compute the rate of change in the shape of the contact region.

Fortunately it is not really necessary to carry out this tiresome program in detail if we are willing to engage in some judicious simplification. First we approximate the leucocyte as a prismatic wheel of diameter D_{cell} attached to the wall of a channel of diameter D_{chan}. This is helpful because the boundary layer is decomposed into two disjoint and uniform segments: the leading boundary layer and the trailing boundary layer. We also assume that the rate of attachment at the leading boundary layer is adequate to maintain a macroscopic contact region under the cell regardless of the rolling velocity (i.e., the leading and trailing boundary layers remain well separated). This last assumption could well be an egregious piece of wishful thinking since even if the bonds permit states of steady peeling and attachment, it is not clear how the rates of these processes at the front and rear of the cell can be exactly matched.

The total tangential traction acting on our model cell is now divided into two parts: the "retarding" traction due to bonds in the boundary layer at the trailing edge of the cell and the "forward" traction due to everything else. Generally, we can approximate the forward traction by a sum of a static component and a component proportional to the hydrodynamic traction (Goldman et al., 1967). In other words

$$\tau_f \approx \tau_{f0} - C D_{cell} \mu V_{fluid} / D_{chan} \,, \tag{32a}$$

where C is a numerical constant.

We integrate Eq. (22a) between the free and clamped extremities to compute an expression for the retarding traction produced by bonds in the trailing boundary layer:

$$\tau_r = \int_{-\infty}^{+\infty} \sigma_{tan} ds \approx +T_{fx}(1 + \cos(\theta_{fx})) \tag{32b}$$

where T_{fx} and θ_{fx} are the microscopic tension and angle at the free extremity of the trailing edge.

Next we integrate Eq.(22a) over the entire circumference of the cell to reach a useful conclusion:

$$\tau_r = -\tau_f \ , \tag{33}$$

i.e., there must be global balance of the tangential traction acting on a prismatic shell.

Setting $V_{fluid} = 0$ and $T_{fx} = T_{crit}$, the traction balance at zero flow allows us to compute the static traction:

$$\tau_{f0} = -\beta^{-1} n_{rtot} \ln(1 + n_{stot} K_l) \ .$$

Thus the static traction tending to cause forward rolling is due entirely to the bonds in the boundary layer at the leading edge of the cell.

Finally, if we divide both sides of Eq. (33) by τ_{f0}, we obtain a general expression for the microscopic tension acting on the trailing boundary layer of a rolling cell:

$$\frac{T_{fx}}{T_{crit}} \approx 1 + \frac{C\beta D_{cell}\mu V_{fluid}}{D_{chan} n_{rtot} \ln(1 + n_{stot} K_l)} \ . \tag{34}$$

This result can be directly substituted into Eq. (28) or Eq.(31) to obtain an expression for the rolling velocity in the case where bonding is mediated by ideal bonds:

$$V_{roll} \approx 0.5 k_{rl} \lambda \ln\left(1 + \frac{C\beta D_{cell}\mu V_{fluid}}{D_{chan} n_{rtot} \ln(1 + n_{stot} K_l)}\right) \tag{35}$$

The main testable prediction of Eq. (35) is the logarithmic dependence of rolling velocity on fluid velocity when fluid velocity is high. The equation also predicts the dependence of rolling velocity on the density of receptors, sites, and all the other molecular parameters. It should, however, be recalled that Eq. (35) applies only for the case of ideal bonds; in the case of slip bonds the rolling velocity is predicted to increase in a super-linear fashion at large fluid velocities (see preceding section).

Dembo et al. (1988) found that the data of Atherton and Born (1973) on the rolling of granulocytes under *in vivo* conditions was in good agreement with the logarithmic relation predicted by Eq. (35). Hammer and Apte (1992) extended this result using a detailed stochastic model of cell rolling. The latter study indicated that, in fact, the properties of ideal bonds are very conducive to the occurrence of stable rolling over a range of fluid velocities. In other words, it was found that rolling is intrinsically unstable and that cells tend to "spin off" the substrate unless attachment is mediated by bonds that are very close to ideal bonds. As far as we know, these two studies remain the main support for the notion that ideal bonds actually exist in nature.

Acknowledgement

The author would like to thank Professor Richard Skalak, Director of the Institute for Mechanics and Materials at the University of California, San Diego, for hospitality and support during the preparation of this manuscript. This work was supported by NIH grant No. AI RO1-21002 and by the U. S. Department of Energy.

References

Atherton, A. and G.V.R. Born, Quantitative investigations of the adhesiveness of circulating polymorphonuclear leucocytes to blood vessel walls, *J. Physiol.* **222** (1972), 447-474..

Atherton, A. and G.V.R. Born, Relationship between the velocity of rolling granulocytes and that of the blood flow in venules, *J. Physiol.* **233** (1973), 157-165.

Bell, G.I., M. Dembo, and P. Bongrand, Cell adhesion: Competition between nonspecific repulsion and specific bonding, *Biophys. J.* **45** (1984), 1051-1064.

Bongrand, P. and G.I. Bell, Cell-cell adhesion: Parameters and possible mechanisms, in *Cell Surface Dynamics: Concepts and Models,* A.S. Perelson, C. DeLisi, and F.W. Weigel, eds., Marcel Dekker, New York, 1984, pp. 459-493.

Dembo, M., and G.I. Bell, The thermodynamics of cell adhesion, *Curr. Topics in Membranes and Transport* **29** (1987), 71-89.

Dembo, M., D.C. Torney, K. Saxman, and D. Hammer, The reaction limited kinetics of membrane to surface adhesion and detachment, *Proc. R. Soc. Lond. B.* **234** (1988), 55-83.

Evans, E.A., and R. Skalak, *Mechanics and Thermodynamics of Biomembranes,* CRC Press, Boca Raton, Florida, 1980, pp. 254.

Evans, E.A., Detailed mechanics of membrane-membrane adhesion and separation. I. Continuum of molecular cross-bridges, *Biophys. J.* **48** (1985), 175-183.

Evans, E.A., D. Burk, A. Leung, and N. Mohandas, Detachment of agglutinin-bonded red blood cells. II. Mechanical energies to separate large contact areas. *Biophys. J.* **59** (1991), 849-860.

Evans, E.A., Physical actions in biological adhesion, in *Biophysics Handbook: Membranes, Vol. I*, R. Lipowsky and E. Sackmann, eds., Elsevier, Amsterdam (in press, 1993).

Goldman, A.J., R.G. Cox, and H. Brenner, Slow viscous motion of a sphere parallel to a plane wall, II. Couette flow, *Chem. Eng. Sci.* **22** (1967), 653-660.

Hammer, D.A. and D.A. Lauffenburger, A dynamical model for receptor-mediated cell adhesion to surfaces, *Biophys. J.* **52** (1987), 475-487.

Hammer, D.A. and S.A. Apte, Simulation of cell rolling and adhesion on surfaces in shear flow: General results and analysis of selectin-mediated neutrophil adhesion, *Biophys. J.* **63** (1992), 35-57.

Krenceski, M.A., J.F. Johnson, and S.C. Temin, Chemical and physical factors affecting performance of pressure-sensitive adhesives, *J.M.S. Rev. Macromol. Chem. Phys.* **C26(1)** (1986), 143-182.

Kuo, S.C. and D.A. Lauffenburger, Relationship between receptor/ligand binding affinity and adhesion strength, *Biophys. J.* (in press, 1993).

Lawrence, M.B. and T.A. Springer, Leucocytes roll on a selectin at physiological flow rates: Distinction from and prerequisite for adhesion through integrins, *Cell* **65** (1991), 859-873.

Springer, T.A., Adhesion receptors of the immune system, *Nature* **346** (1990), 425-433.

Springer, T.A. and L.A. Lasky, Sticky sugars for selectins, *Nature* **349** (1991), 196-197.

Takeichi, M., Cadherin cell adhesion receptors as a morphogenic regulator, *Science* **251** (1991), 1451-1455.

Zhu, C., A thermodynamic and biomechanical theory of cell adhesion, *J. Theo. Biol.* **150** (1991), 27-50.

MICAH DEMBO
THEORETICAL BIOLOGY AND BIOPHYSICS GROUP
THEORETICAL DIVISION
LOS ALAMOS NATIONAL LABORATORY
LOS ALAMOS, NEW MEXICO 87545 U.S.A.
MXD@ORGANELLE.LANL.GOV

Lectures on Mathematics in the Life Sciences
Volume **24**, 1994

Toward Quantifying the Activation of Helper T Lymphocytes by Antigen Presenting Cells

JENNIFER J. LINDERMAN,

NANCY G. BERRY, AND DEBRA F. SINGER

April 21, 1993

ABSTRACT. Helper T lymphocytes (Th cells) respond to antigen presented by B lymphocytes and macrophages (antigen-presenting cells, APC) in the context of MHC class II molecules via a process termed antigen presentation. We present a mathematical framework to begin to quantify the relationship between antigen concentration in solution and subsequent Th cell activation and to explain our single cell experiments. First, we develop models of antigen uptake and processing by APC. We find that antigen uptake via non-specific fluid phase endocytosis or, when applicable, receptor-mediated endocytosis, is sufficient to allow the presentation of 100–2000 MHC-antigen complexes on the APC. We use this model to explain the enhancement of antigen presentation with receptor-mediated uptake in two different systems and to explain observed competition between peptides binding to the same MHC. Second, we develop simple models of the interaction and adhesion of APC and Th cells. We find that stochastic effects may suggest an explanation for the failure of some Th cells to respond to APC. We further estimate the number of bound T cell receptors as a function of time and after a prolonged period of cell-cell contact. These and future generation models of the steps involved in Th cell activation may be able to suggest methods by which the immune response may be manipulated at the level of antigen presentation.

1991 *Mathematics Subject Classification.* Primary 92C45,92-06; Secondary 78A70.

Supported by funding from the NSF Presidential Young Investigator program, the Whitaker Foundation, and a Cindy Yoder Research Award.

The authors wish to thank Dr. Rod Nairn for the gift of cell lines and Ms. Saroja Ramanujan for preliminary model calculations.

This paper is in final form and no version of it will be submitted for publication elsewhere.

1. Introduction

The ability to manipulate, *i.e.* enhance or suppress, the immune response is of great medical value. To do this precisely and specifically, one needs to first understand the quantitative and kinetic relationships between the different components of the immune system and their activation. In our work, we have focused on the interaction between antigen-presenting cells and T helper lymphocytes and the ensuing activation of those T helper cells. We focus on these events, collectively termed antigen presentation, because they mark an early and critical step in the development of an immune response. In particular, activated T helper cells are responsible for "helping" other cells of the immune system, leading to, among other things, increased populations of cytotoxic T lymphocytes and antibody-producing B lymphocytes. This help is typically delivered via direct cell-cell contact and the secretion of lymphokines. The actions of cytotoxic T cells, which kill antigen-infected cells, and of antibodies, which bind and inactivate soluble antigen, are crucial to the immune system's ultimate goal: the elimination of foreign antigen from the body.

The immune system is poised quite precariously. It must respond rapidly and strongly— but not so strongly as to endanger the organism with the side effects of an overzealous response—to small doses of life-threatening antigens. It should not respond at all to "self" antigens (*i.e.*, molecules that are derived from the organism itself), even though these "self" antigens may very closely resemble a foreign antigen. Much recent research has been directed at understanding when the immune system fails to keep this balance. A primary example of this failure occurs in autoimmune diseases, which occur when the organism mounts a debilitating and often life-threatening reaction to its own antigens. Approximately 5–7% of the population is affected by some type of autoimmune disease. Examples of autoimmune diseases in which T helper cell activation is triggered in error include insulin-dependent diabetes mellitus, multiple sclerosis, hyperthyroidism, rheumatoid arthritis, myasthenia gravis, and systemic lupus erythematosus (1).

Although aspects of T helper cell activation have been investigated in numerous studies, most descriptions of the events that occur during antigen presentation are qualitative and anecdotal. A further complication is the fact that some of the key variables thought to control activation, for example, the number of occupied T cell receptors on the T helper cell, are at present unmeasurable. With this backdrop, we have found mathematical modeling to be useful in interpreting literature data and our own data and in suggesting mechanisms behind T helper cell activation. Ultimately, we hope that our modeling approach will provide an informed basis by which to guide both future studies of antigen presentation and clinical manipulation of the immune response at the level of antigen presentation.

We include in this paper a brief summary of the events of antigen presentation and a description of a single cell experiment to give more insight into the process. We then turn to attempts to model the events of antigen presentation.

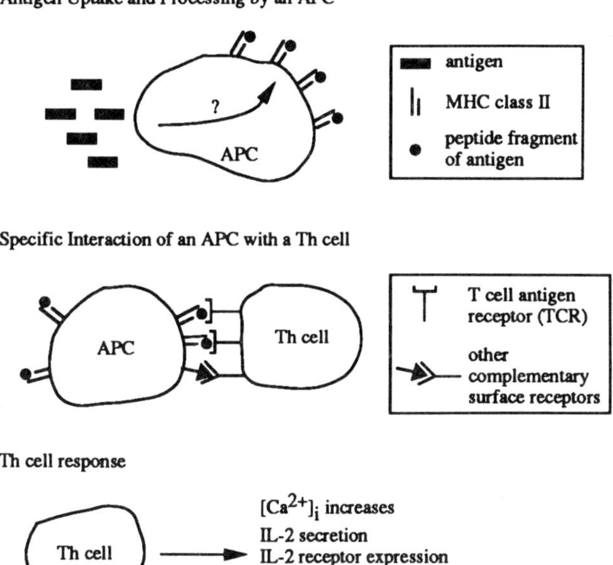

FIGURE 1. Schematic of the events of antigen presentation.

2. Background

There are three steps that occur in antigen presentation (Fig. 1). First, antigen-presenting cells (APC) must prepare or process antigen for recognition by T helper lymphocytes (Th cells). Second, the APC must then interact with the Th cell to form a cell-cell conjugate. Third, the Th cell, as a result of its interaction with the APC, is activated. A brief description of each of these steps is given below; more details are included in the modeling section of this paper as warranted.

2.1. Antigen uptake and processing by the APC. Th cells cannot be activated by soluble antigen. APC, typically B lymphocytes, macrophages, or dendritic cells, must first internalize and process the antigen and then display the relevant antigenic fragment on their membrane surface. It is this form of the antigen that is recognized by the Th cell. Because of difficulties in working with dendritic cells, most reported studies of APC use B cells and macrophages; we will focus on those studies.

To summarize briefly, APC are believed to perform the following steps in the processing of protein antigens (2-4). Protein antigens are internalized non-specifically (via fluid phase pinocytosis) or specifically (via receptor-mediated endocytosis). Internalized antigens are delivered to acidic intracellular processing compartments where enzymes such as cathepsin D break down the protein into smaller peptide fragments. Specialized glycoproteins termed MHC class II molecules are present in the membranes of these acidic compartments and can bind to appropriate peptide fragments of approximately 8–20 amino acids in length (5,6). The identity of the peptide and the genotype of the cell (and thus the structure of the MHC molecule) determine whether binding can occur. It is also known that one particular MHC molecule can bind many different peptides (7); in loose terms, the binding is somewhat sloppy, but this allows the presentation of many different peptides without the need for an equally variant population of MHC in the body. Following binding of MHC to peptide antigen, this complex is returned to the APC cell surface. The net result of these steps, which the APC accomplishes independently of any interaction with the Th cell, is the display of MHC class II-peptide antigen (hereafter termed MHC-ag) complexes on the surface of the APC. For antigen concentrations of 0.005–1 μM (the lower end of the range requiring the efficient internalization of antigen via receptor-mediated endocytosis), sufficient complexes are displayed after roughly one hour (macrophages) and 6–8 hours (B cells) to allow activation of appropriate Th cells (8-10).

2.2. Specific interaction of an APC with a Th cell. Th cells possess T cell antigen receptors (TCR), cell surface molecules that have been shown to simultaneously recognize both the MHC class II molecule and the antigen fragment it binds (11,12). The interaction of an APC with a Th cell is specific: only if the APC displays MHC-ag complexes and the Th cell the appropriate TCR will the two cells form a stable contact area and Th cell activation occur.

In addition, several other molecules on the membranes of the APC and Th cell are now known to participate in the adhesion between the two cells, although these interactions are not specific to the antigen that is being presented (13,14). Some of the key molecular events that are known to occur in the development of the adhesion include binding of LFA-1 on the Th cell to ICAM-1 on the APC, binding of CD2 on the Th cell to LFA-3 on the APC, and binding of CD4 on the Th cell to the MHC class II molecule on the APC. Evidence has been reported showing that the affinity of LFA-1 on the Th cell for its ligand ICAM-1 is enhanced within about 5 min and for a period of about 30 min or more after TCR bind specifically to their appropriate MHC-ag ligand (15). A reasonable interpretation of this result, together with other data, is that when APC and Th cells first meet, a few MHC-ag/TCR bonds are formed and that these provide the positive signal to increase the LFA-1 affinity. There is limited evidence that the adhesion of APC and Th cell may recruit additional molecules, perhaps by diffusion, into the contact area: TCR and CD4 redistribution has been noted by Kupfer and coworkers, who examine the distribution qualitatively on single fixed cells by using fluorescent antibodies directed against these molecules (16,17).

It is believed that the Th cell response depends on the "signal" it receives from its interaction with the APC. In particular, the most likely interpretation of a host of data (*e.g.*, 18) reporting an increase in Th cell activation upon increases in the concentration of antigen with which APCs are incubated is that the Th cell response must be a function of the number of TCR bound to MHC-ag complexes. There are currently no data available on the number of bound TCR or bound "adhesion" molecules (*e.g.*, LFA-1) that result from the APC/Th cell interaction, due to the failure of current biochemical and immunological techniques to allow this measurement. Because this information is crucial to understanding the Th cell response, new techniques need to be developed to measure or predict this quantity.

2.3. Th cell response. The Th cell, as a result of its interaction with an APC displaying the appropriate MHC-ag complexes, is activated. Several distinct events have been noted in this response: the Th cell secretes the lymphokine interleukin-2 (IL-2), expresses IL-2 receptors, and, upon binding of IL-2 to its own IL-2 receptors, proliferates. To assay Th cell activation, one common method is to assay for IL-2 secretion by collecting the medium obtained from the Th cell culture and incubating it with an IL-2 dependent cell line. The growth in this cell line is then assayed by ^{3}H-thymidine uptake. Because of the indirect and delayed nature of this assay, it is difficult to relate this response quantitatively with the APC/Th cell interaction.

More recently, a very early event in Th cell stimulation, an increase in the cytosolic free calcium concentration, has also been measured (19-21). Increases in cytosolic calcium have been noted to be a common result of the activation of the inositol phosphate signal transduction pathway in many cell types. The general steps of this pathway are the topic of many reviews (*e.g.*, 22,23) and include receptor-ligand binding, activation of the enzyme phospholipase C, cleavage of phosphatidylinositol 4,5-bisphosphate to produce inositol 1,4,5-trisphosphate (IP_3), binding of IP_3 to sites on the membrane of intracellular calcium stores, and opening of calcium channels on the store membrane to release calcium into the cytosol. Changes in the concentration of intracellular calcium in populations of cells or single cells can be monitored using fluorescent calcium-sensitive dyes and can be measured in single cells by using digital fluorescence imaging techniques (24,25). Activation of other signal transduction pathways in the Th cell have also been implicated in describing the full Th cell response, and it is likely that more than one of these pathways must be utilized for complete activation of primary Th cells (26-28).

2.4. Quantitative and kinetic relationships between key variables. The three distinct steps of antigen presentation involve the following unknown relationships between fundamental parameters of the system. Uptake and processing of antigen by an APC determines the relationship between the antigen concentration in solution and the number of MHC-ag complexes displayed at any given time (Fig. 2a). Interaction of an APC and Th cell presumably results in the binding of TCR to an extent determined partly by the number of MHC-ag complexes available (Fig. 2b). Other cell surface molecules on the Th cell may

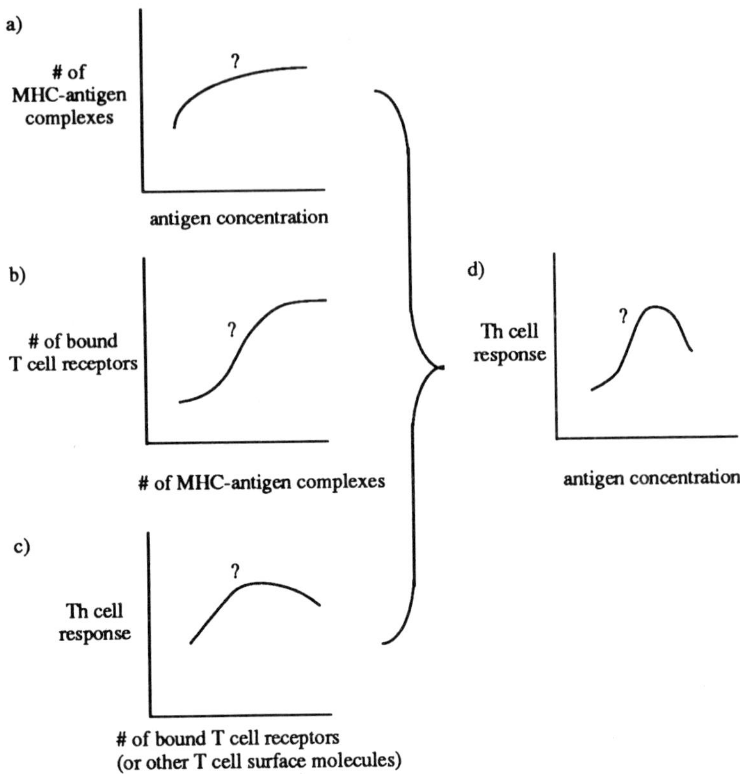

FIGURE 2. Critical relationships for understanding antigen presentation.

also be bound. Finally, the number of bound TCR, and most likely the number of other bound Th cell surface molecules as well, determines the Th cell response (Fig. 2c). The net result of the three relationships just described is a relationship between antigen concentration and a Th cell response (Fig. 2d). Although this relationship can be assayed using many cells in suspension (as mentioned briefly in Section 2.3), our models and single cell experiments aim to look at the relationships first at the single cell level. In this paper, we discuss models focused on addressing the relationships shown in Figs. 2a-c and single cell experiments focused on discovering the relationship shown in Fig. 2d.

3. Experimental materials and methods

The murine B cell lymphoma line TA3 (used as APC) and the murine Th hybridoma cell line 3A9 were gifts from Dr. Rod Nairn and were cultured as previously described (29). Hen egg lysozyme (Sigma) was used as an antigen.

APC were incubated with 10 μM antigen overnight to allow for processing and display of MHC-ag complexes. Immediately before an experiment, Th cells were loaded with calcium-sensitive dye by a 30 min incubation at 37 °C in a

5% CO_2 atmosphere with 5 μM fura-2 acetoxymethyl ester (Molecular Probes). Th cells were then rinsed several times with buffer (RPMI-1640 supplemented with 1.0% antibiotic antimycotic solution, 20 mM HEPES, 0.6 mM Ca^{2+}). Single poly-D-lysine coated glass coverslips were mounted in a home-built temperature-controlled flow chamber placed on the stage of an inverted microscope. Th cells were added and allowed to adhere to the coverslip. After rinsing away unattached Th cells, APC were added to the flow chamber so that individual APC could contact Th cells and form cell-cell couples.

Cells were visualized using a Nikon Diaphot inverted microscope equipped with a Hg arc lamp, 40x NA 1.3 oil objective, 334 and 365 band pass filters, 400 nm dichroic mirror, 400 nm long pass filter, and Nikon Nomarski DIC TMD-NT2 optics. A neutral density filter was used to minimize fura-2 bleaching and cell injury. Images were collected using a charge-coupled device (CCD) camera (Photometrics Ltd.). Nomarski and fluorescence images were taken alternately at intervals of several seconds. ISee imaging software (Inovision Corp.) running on a Sun 4/330 computer was used to control filter wheel, shutter, and camera operation as well as to perform subsequent image processing and analysis.

Excitation of the dye fura-2 was accomplished at 334 nm and 365 nm in order to capture two images which can be ratioed. Calibration images were taken by imaging three solutions in the flow chamber positioned on the microscope stage: 10 μM fura-2 free acid (Molecular Probes) in phosphate-buffered saline supplemented with 2 mM $CaCl_2$, 10 μM fura-2 free acid supplemented with 2 mM EDTA, and RPMI alone (for background subtraction). The intracellular calcium concentration $[Ca^{2+}]_i$ was calculated from

$$[Ca^{2+}]_i = K_d\beta\left(\frac{R - R_{min}}{R_{max} - R}\right) \tag{3.1}$$

where $R = F_{334}/F_{365}$, the fluorescence intensity of a cell image at 334 nm excitation divided by the fluorescence intensity of a cell image at 365 nm excitation; $R_{min} = F_{334}/F_{365}$ for a solution of fura-2 with no calcium; $R_{max} = F_{334}/F_{365}$ for a solution of fura-2 with 2 mM calcium; $\beta = F_{365}(\text{no calcium})/F_{365}(2\text{ mM calcium})$; $K_d = 224$ nM (30).

4. Experimental results

Our experiments are aimed at observing the interaction of live APC and Th cells at the single cell level. We use as a model system one that has been well-characterized by other investigators, the murine B cell lymphoma line TA3, the antigen hen egg lysozyme, and the murine Th hybridoma cell line 3A9 (29, 31). The 3A9 cells are known to be specific for the hen egg lysozyme fragment 46-61. We label Th cells with a fluorescent calcium-sensitive dye which serves two purposes: (1) it allows us to distinguish fluorescent Th cells from non-fluorescent APC in the same field of view, and (2) it allows us to follow changes in intracellular free calcium concentration $[Ca^{2+}]_i$ in Th cells.

The results of one such experiment are shown in Fig. 3. Shown are 9 frames out of a time-lapse sequence of several hundred images. The fluorescence images are not shown; rather, the Nomarski images taken alternately with fluorescence images are shown. On the Nomarski images, both B and Th cells are visible. The Th and B cells of interest in this sequence are labeled. The B and Th cell surface areas are estimated at 400–500 μm^2. In Fig. 4, $[Ca^{2+}]_i$ in the Th cell of Fig. 3 is shown as a function of time.

In this particular experiment, a Th cell was found very close to a B cell at t=737 sec. A short time later (811 sec), the Th cell extended a pseudopod and contacted the B cell. The interaction was brief (no interaction was seen by 885 sec), only a small contact area (estimated at 10 μm^2) was formed between the B and Th cells, and $[Ca^{2+}]_i$ remained essentially constant in the Th cell. Because no significant increase in $[Ca^{2+}]_i$ in the Th cell was seen, we classify this interaction as one which did not produce a Th cell response.

Later in the same experiment (1183 sec), several B cells dropped from solution onto the coverslip near the same Th cell. The Th cell then interacted with these B cells, forming an extensive contact area (estimated at 50 μm^2) with one of the B cells by 1407 sec. Note in Fig. 4 that shortly after the Th cell interacts with this second B cell, the Th cell responds with approximately a ten-fold increase in $[Ca^{2+}]_i$. At times 2241, 2687, and 3060 sec, the flow chamber was rinsed with buffer; the APC/Th cell conjugate was undisturbed by the fluid flow.

We find in general that not all Th cells contacting APC respond with increases in $[Ca^{2+}]_i$. Although one possibility for this is that only some Th cells are capable of responding at all, the rare experiment shown here, in which a Th cell contacted two B cells successively and responded only to the second one, suggests that this may not be the case. Preliminary data suggest that the percentage of Th cells responding is an increasing function of antigen dose (the concentration of antigen with which APC are incubated).

5. Modeling approaches to the understanding of Th cell activation

With this brief background, we now turn to the use of mathematical models to begin to quantify the events of antigen presentation. In particular, we aim to discover the relationship between antigen concentration and MHC-ag display by an APC (Fig. 2a) and the number of bound TCR on a Th cell interacting with an APC (Fig. 2b). We only begin to address the relationship between bound TCR and a Th cell response (Fig. 2c), in an attempt to discover why some Th cells but not others respond in our experimental system.

5.1. Calculation of the number of MHC-ag complexes on the APC. In order to quantitatively understand the relationship between antigen dose and subsequent Th cell activation (Fig. 2d), it is necessary to first understand the relationship between antigen dose and the number of MHC-ag complexes dis-

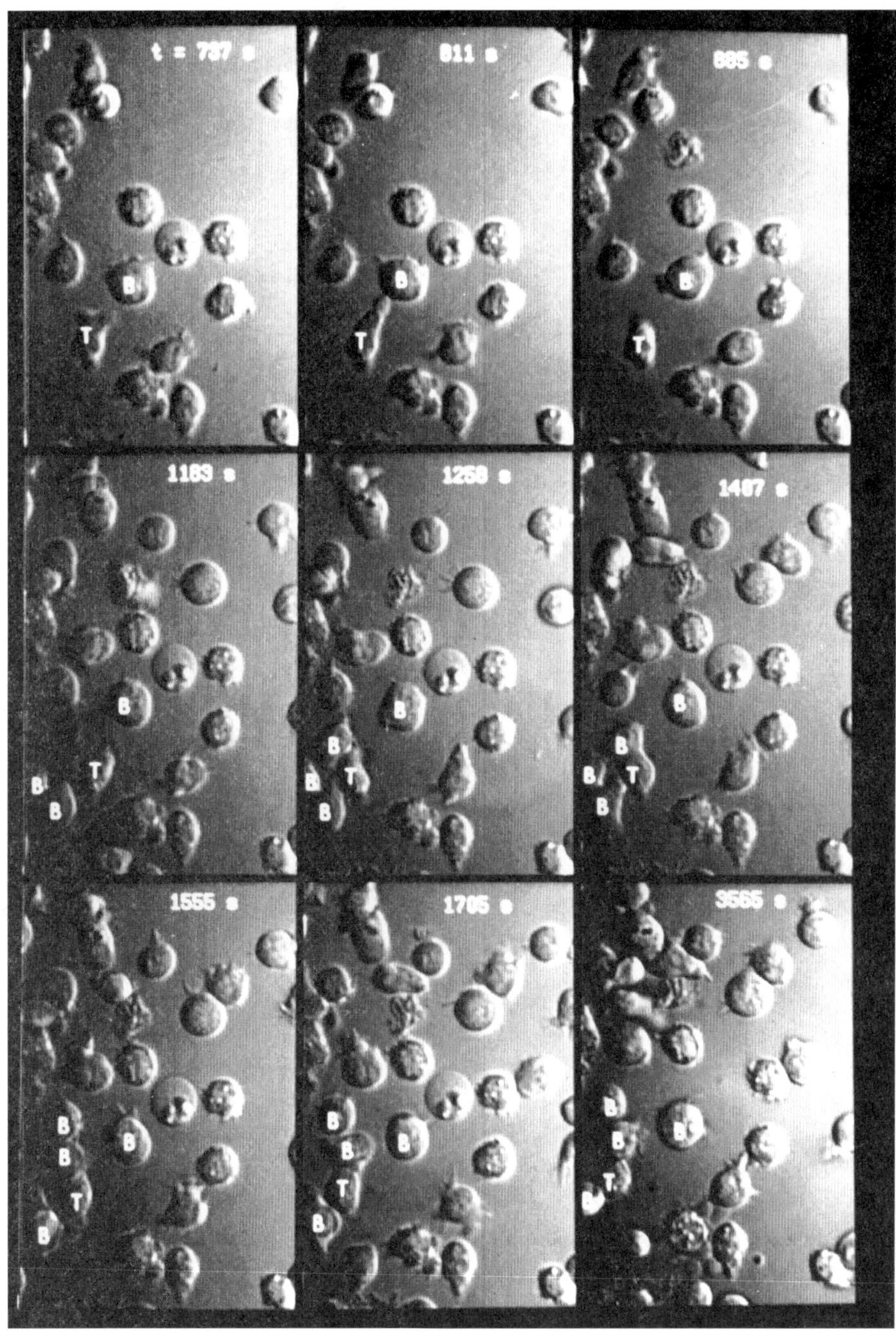

FIGURE 3. Interaction of single Th and B cells. Nomarski images at nine different times are shown.

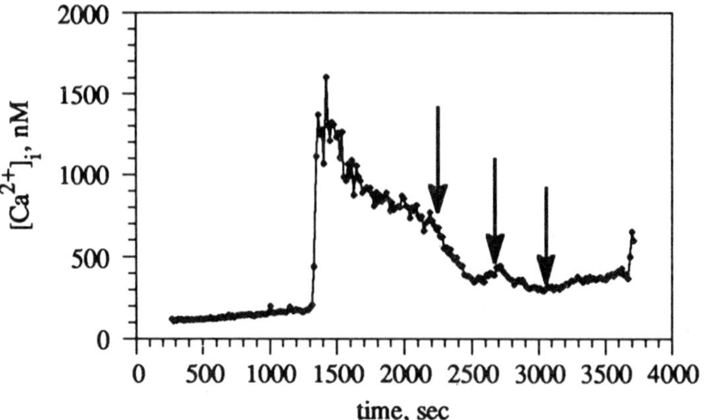

FIGURE 4. Calcium response of Th cell shown in Fig. 3. $[Ca^{2+}]_i$ remains approximately constant throughout an interaction with a first B cell but rises sharply following the interaction with a second B cell. Arrows mark times at which the flow chamber was rinsed with buffer.

played to the Th cell (Fig. 2a). The number of MHC-ag complexes necessary on the surface of the APC is believed to be very small, on the order of 200–2000, as determined from assays several purification steps removed from the true physical situation or from model systems (12,32-34). This quantity is exceptionally difficult to measure accurately.

Prediction of the number of MHC-ag complexes is difficult as well. Antigen may enter the cell through either a non-specific or a receptor-mediated route, and evidence suggests that antigen internalized via either route is delivered to endosomes and perhaps pre-lysosomes and lysosomes. In addition, there is likely competition for MHC binding from other antigens and even serum proteins present in most standard assays.

5.1.1. *Receptor-mediated versus nonspecific uptake of antigen.* In order to predict the number of MHC-ag complexes on the surface of the APC as a function of time and antigen concentration, we model the events of uptake and processing in a single cell according to the schematic shown in Fig. 5 (35). Protein antigens for which there is a corresponding surface receptor may bind that receptor with association rate constant k_f^* ($M^{-1}min^{-1}$) and dissociate from the receptor with rate constant k_r^* (1/min). Unbound cell surface MHC molecules are assumed to be internalized with rate constant k_v (1/min). Antigen may be taken into the cell non-specifically (unbound to a receptor) with rate constant k_u (μm^3/min) which can be related to the rate constant k_v by

$$k_v = \frac{k_u(SA_v)}{V_v(SA_c)} \tag{5.1}$$

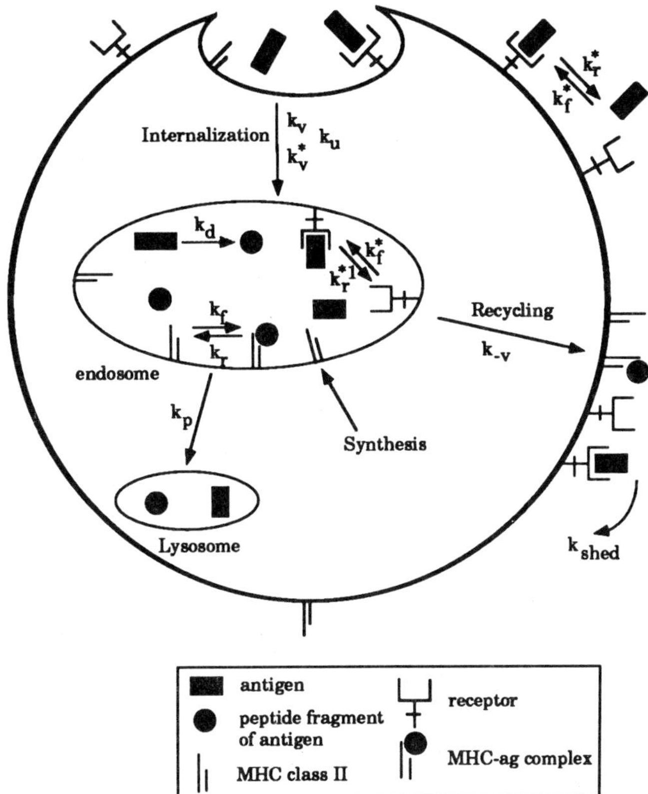

FIGURE 5. Model schematic for the processing of antigen by an APC.

where V_v is the endosomal volume per cell, SA_c is the surface area of the cell, and SA_v is the endosomal surface area. Antigen bound to surface receptors may be internalized with rate constant k_v^* (1/min). All antigens are assumed to be delivered to and processed in endosomes, although some recent data suggest that the site of antigen processing, at least for some antigens, may be further along in the internalization pathway (36).

Receptor-bound antigen found in endosomes can dissociate from its receptor with rate constant k_r^{*1} (1/min), which may be greater than k_r^* due to the low pH of the endosome. This antigen may also rebind the receptor; it is assumed that the rebinding rate constant k_f^* is identical to that at the cell surface. Unbound antigen is degraded by proteases into its relevant peptide fragment with a rate constant k_d (1/min). Peptides may then bind with rate constant k_f ($M^{-1}min^{-1}$) to MHC molecules found in the endosome and may dissociate from MHC molecules with rate constant k_r (1/min). For the few peptide/MHC systems for which values of k_f and k_r have been measured, little variation in these values with the identity of the peptide and MHC species has been reported

(37) and therefore we assume the values to be similar for all systems. Such an approximation can be updated once more complete information becomes available.

MHC-ag complexes, free MHC molecules, and receptors are assumed to be recycled back to the cell surface with rate constant k_{-v} (1/min). Undegraded unbound antigen and unbound peptides are assumed to be shuttled to lysosomes with rate constant k_p (1/min) for further degradation. Newly synthesized MHC molecules are assumed to be delivered to endosomes. We note that newly synthesized MHC molecules are believed to be associated with another molecule, the invariant chain, which may block the binding site until the MHC molecule reaches the endosome (4). We neglect any details of invariant chain dissociation in this model. Finally, we assume that MHC-ag complexes are lost from the cell surface with rate constant k_{shed} (1/min). This loss may represent reuptake and routing of the complex to the lysosome or loss via any other pathway.

We assume that the total number of receptors and MHC molecules does not change, at least for the time scale of this analysis, and thus write the conservation equations:

$$A_0 = A + B + C + C^* \tag{5.2}$$

$$R_0 = R + (RL) + R^* + (RL)^* \tag{5.3}$$

where A_0 is the total number of MHC molecules per cell, A is the number of free MHC molecules on the cell surface per cell, B is the number of free MHC molecules in endosomes per cell, C is the number of MHC-ag complexes in endosomes per cell, C^* is the number of MHC-ag complexes on the cell surface per cell, R_0 is the total number of receptor molecules per cell, R is the number of free receptors on the cell surface per cell, (RL) is the number of receptor-antigen complexes on the cell surface per cell, R^* is the number of free receptors in endosomes per cell, and $(RL)^*$ is the number of receptor-antigen complexes in endosomes per cell.

The equations describing the time course of MHC-ag formation are then:

$$\frac{dR}{dt} = -k_f{}^* R L_0 + k_r{}^*(RL) + k_{-v} R^* \tag{5.4}$$

$$\frac{d(RL)}{dt} = k_f{}^* R L_0 - k_r{}^*(RL) - k_v{}^*(RL) + k_{-v}(R_0 - R - (RL) - R^*) \tag{5.5}$$

$$\frac{dR^*}{dt} = -k_{-v} R^* + k_r{}^{*1}(R_0 - R - (RL) - R^*) - k_f^* R^* L^* \tag{5.6}$$

$$
\begin{aligned}
\frac{dL^*}{dt} N V_v N_A &= \frac{k_v V_v (SA_c)}{(SA_v)} L_0 N_A + k_r{}^{*1}(R_0 - R - (RL) - R^*) \\
&\quad - k_f^* R^* L^* - (k_d + k_p) L^* V_v N N_A
\end{aligned} \tag{5.7}
$$

$$
\frac{dA}{dt} = k_{-v}(A_0 - A - C - C^*) - k_v A - k_{shed} A \tag{5.8}
$$

$$
\frac{dC}{dt} = k_f (A_0 - A - C - C^*)(Ag) - (k_r + k_{-v}) C \tag{5.9}
$$

$$
\frac{dC^*}{dt} = k_{-v} C - k_{shed} C^* \tag{5.10}
$$

$$
\begin{aligned}
\frac{d(Ag)}{dt} N V_v N_A &= k_d L^* N V_v N_A + k_r C - k_f (A_0 - A - C - C^*)(Ag) \\
&\quad - k_p (Ag) N V_v N_A
\end{aligned} \tag{5.11}
$$

where we have used Eqns. (5.2,5.3) to eliminate the variables B and $(RL)^*$. L_0 is the extracellular antigen concentration in M (assumed constant), L^* is endosomal antigen concentration in M, and (Ag) is the concentration of peptide fragments of the original antigen. N is the number of endosomes per cell and N_A is Avogadro's number. Note that the fluid phase pinocytosis route is always functioning but that in the absence of receptors specific for antigen, Eqns. (5.4–5.6) can be neglected. Parameter estimates from a variety of literature data can be found in Singer and Linderman (35). Model equations were solved numerically.

We can use our model to predict the number of MHC-ag complexes expressed on the surface of an APC and to compare the efficiency of complex expression when antigen uptake is via the fluid phase pinocytosis (FPP) or receptor-mediated endocytosis (RME) pathway. For this calculation, we focus on two recent reports in which the Th cell response of IL-2 secretion was measured for a variety of antigen concentrations and for antigen internalized through the FPP or RME pathway. Casten and Pierce (10) used TPc9.1 T cells, murine splenic B cells, and the antigen pigeon cytochrome c. They examined the uptake of both pigeon cytochrome c and pigeon cytochrome c coupled to antibodies to cell surface immunoglobulin (acting as receptors). Their data show that approximately 50% and 100% Th cell stimulation occur at antigen concentrations of about 0.003 and 0.1 μM for RME and about 1.0 and 4.0 μM for FPP uptake mechanisms. We suggest that similar levels of stimulation result from the display of similar numbers of MHC-ag complexes and test this with model calculations. The results of model simulations for this system are shown in Fig. 6a. Parameter values are given in the figure legend; we note that we have used a larger value for

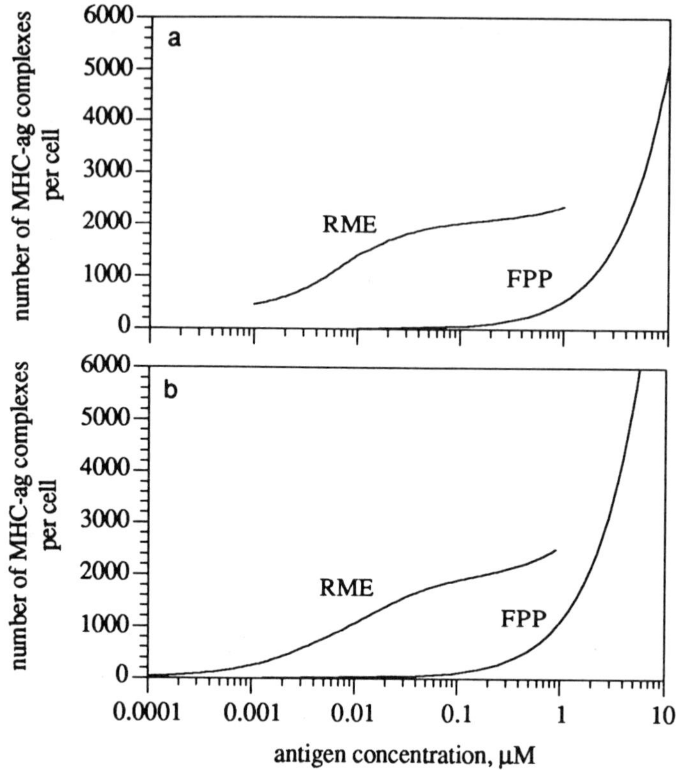

FIGURE 6. Model comparisons for antigen uptake via fluid phase pinocytosis (FPP) or receptor-mediated endocytosis (RME) routes. (a) A comparison to the data of Casten and Pierce (10). Parameter values: k_v=0.009 min^{-1}, k_v^*=0.04 min^{-1}, k_{-v}=0.05 min^{-1}, SA_c = 900 μm^2, SA_v = 1.5 μm^2, V_v = 4.0x10^{-17} l, N=10, A_0=1.1x10^5, R_0 = 6.5x10^5, k_f=250 M^{-1}min^{-1}, k_r=4.2x10^{-4} min^{-1}, k_f^*=4x10^5 M^{-1}min^{-1}, k_r^*=2.7x10^{-3} min^{-1}, k_r^{*1}=0.27 min^{-1}, k_d=0.012 min^{-1}, k_p=0.105 min^{-1}, k_{shed}=3.5x10^{-3} min^{-1}, time = 8 hours. (b) A comparison to the data of Chu *et al.* (39). Parameter values: k_v=0.03 min^{-1}, k_v^*=1.5 min^{-1}, k_{-v}=0.05 min^{-1}, SA_c = 2200 μm^2, SA_v = 1.5 μm^2, V_v = 4.0x10^{-17} l, N=10, A_0=1.1x10^5, R_0 = 4x10^5, k_f=250 M^{-1}min^{-1}, k_r=4.2x10^{-4} min^{-1}, k_f^*=1x10^6 M^{-1}min^{-1}, k_r^*=0.002 min^{-1}, k_r^{*1}=6 min^{-1}, k_d=0.005 min^{-1}, k_p=0.105 min^{-1}, k_{shed}=5x10^{-4} min^{-1}, time = 2 hours.

k_{shed} than in earlier discussions of this model (35) because recent data suggest that MHC-ag complexes have a relatively short half-life of 2–4 hours in this particular system (38). Note that reasonable agreement with the quantitative data of Casten and Pierce (10) is found. In addition, we estimate that numbers of MHC-ag complexes necessary for 50% and maximal Th cell stimulation in their system are approximately 600 and 2000, respectively.

For a comparison of the quantitative advantage conferred by receptor-mediated internalization in another system, we turn to the data of Chu and Pizzo (39). These investigators used mouse peritoneal macrophages, the 3A9 T cell hybridoma line, and the antigen hen egg lysozyme. The antigen was internalized either by FPP or, when complexed with α_2-macroglobulin, via α_2-macroglobulin receptors and RME. The results of model simulations for this system are shown in Fig. 6b. In this case, parameter values typical of macrophages are used. In addition, a small value for k_{shed} is used because of suggestions in many systems that the half-life of MHC-ag complexes is on the order of 10 or more hours (40,41). Chu and Pizzo find similar levels of IL-2 secretion for antigen concentrations of about 0.002 μM (RME) and 0.6 μM (FPP); these antigen concentrations give a threshold, or barely detectable, response in their system. In addition, they find similar levels of IL-2 secretion for antigen concentrations of about 0.03 μM (RME) and 2 μM (FPP). Note that reasonable agreement with these quantitative data is found in the model results shown in Fig. 6b. Further, we can estimate that approximately 500 complexes are sufficient for threshold stimulation in their system.

Recent estimates of the minimal number of MHC-ag complexes needed to stimulate a detectable Th cell response are on the order of a few hundred complexes (33,34). Thus our model is consistent with these data and has the added advantage of predicting the dependence of this number on key physiological parameters.

5.1.2. *Competition for MHC binding among different peptides.* The analysis presented above can easily be extended to the situation when multiple peptides are present that can compete for binding to the same MHC molecule. In model systems, these peptides are presumably those to which the particular Th cell type used will not respond. The competing peptides may be added as proteins or as peptides, they may be secreted by the APC or other cells present, or they may be present in the medium used to grow the cells. *In vivo*, these competing peptides are likely derived from other antigens to which the body should respond and may very well be able to interact with the many different TCR available on different Th cells. Competition between different peptides for MHC binding has been observed qualitatively (42-45), and the effect on the number of MHC-ag complexes available for Th cell interaction can be quantified with our model.

The model presented schematically in Fig. 5 can be applied to the case of competition among peptides by simply adding equations to account for the gen-

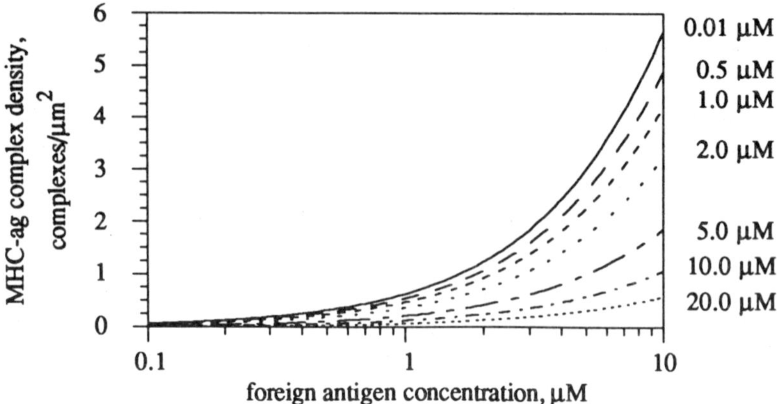

FIGURE 7. Predicted density of stimulatory MHC-ag complexes on the surface of a macrophage APC after 1 hour. The concentration of foreign antigen, the antigen that gives rise to immunogenic MHC-ag complexes, is given on the abscissa. Simulations were done for 0.01-20 μM of competing proteins. Model equations and parameter values are given in Singer and Linderman (46).

eration of MHC-ag complexes of different types (46). Sample results of such analyses are shown in Fig. 7 for a macrophage APC.

A test of the model's ability to describe the behavior of APC can be made by a comparison with the data of Lorenz *et al.* (45). Lorenz and co-workers addressed the sensitivity of macrophage APC to changes in competing protein concentrations by varying the concentration of serum in their assay medium. Simulation results showing MHC-ag complex density for a fixed stimulatory antigen concentration of 1.0 μM as a function of the concentration of competing (non-stimulatory) protein are shown in Fig. 8. Our simulations are consistent with Lorenz and co-workers' data, which show a 50% inhibition of the Th cell response at a competing protein concentration of 5.0 μM as compared to the response with no competing protein present.

5.2. The initial contact between an APC and a Th cell. As described earlier, our single cell experiments demonstrate that not all Th cells which contact APC displaying appropriate MHC-ag complexes respond with an increase in $[Ca^{2+}]_i$. In fact, in the micrographs shown in Fig. 3, the Th cell does not respond to the first APC it contacts, but only to the second. While it is indeed rare for us to observe the interaction of a particular Th cell with two different B cells during the course of a single experiment, we often observe Th cells which contact B cells but do not respond.

There are several possible explanations for the failure of a Th cell to respond

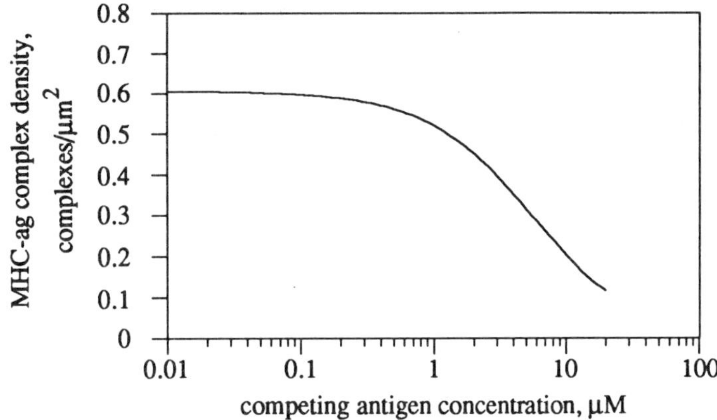

FIGURE 8. Predicted stimulatory MHC-ag complex density (complexes/μm^2) in macrophages. The foreign antigen concentration is fixed at 1.0 μM and varying concentrations of a competing protein are used. Parameter values are given in Singer and Linderman (46).

to one particular APC but not another. For example, one likely possibility is that APCs themselves may be heterogeneous in the number of MHC-ag complexes they display, due perhaps to cell-to-cell variability in rates of internalization of antigen, number of MHC molecules, etc. Alternatively, the Th cell may require multiple interactions with different APC cells in order to be triggered, and we observe only a small subset of these interactions. We know of no reports to support this theory, and we think it unlikely that the Th cells have had that opportunity in our system.

Yet another possibility is suggested by the very small numbers of MHC-ag complexes that are known to be sufficient for Th cell stimulation (calculations above; 33,34). When very small numbers of molecules participate in a reaction (in this case, the binding of MHC-ag to TCR), stochastic effects may come into play. These effects may be quite pronounced in this system, for only a small portion of the Th cell membrane initially comes into contact with the APC surface and thus few MHC-ag complexes are likely to be present in this contact area. For an initial contact area on the order of 10 μm^2 and APC sizes varying from 200–2000 μm^2 depending on the cell type, only 0.5–5% of the total number of MHC-ag complexes are likely to be found in the contact area. Given that there are likely only 200–2000 MHC-ag complexes on an APC, the Th cell may only be able to "see" in the neighborhood of 1–100 MHC-ag complexes during this initial contact.

A reasonable hypothesis suggests that the APC-Th cell conjugate is considered "productive" and is sustained if a threshold number of antigen complexes is recognized on the surface of the APC through MHC-ag binding with the TCR. If this threshold level of complexes is identified, the Th cell and APC remain in contact and the contact area increases in time for a finite period. The Th cell is activated, and the ensuing aspects of an immune response are initiated. However, if the Th cell fails to recognize the threshold level of complexes, the cells dissociate and no response is triggered.

If we assume that the MHC-ag complexes are distributed randomly on the APC surface, then the probability P_c that there are c complexes present in the initial contact area between APC and Th cell is described by a binomial distribution:

$$P_c = \frac{Q!}{(Q-c)!c!} A^c (1-A)^{Q-c} \tag{5.12}$$

where Q is the total number of MHC-ag complexes on the APC surface and A is the fraction of APC surface area that the Th cell contacts. Ignoring for now the details of TCR/MHC-ag binding, if we assume that the Th cell must "see" some threshold number c_{thresh} of MHC-ag complexes in order to respond, then the probability P that the Th cell responds is given by:

$$P = \sum_{c=c_{thresh}}^{Q} P_c \tag{5.13}$$

A plot of the probability of Th cell response P versus the expected mean number of complexes in the contact area, QA, for different values of c_{thresh} is shown in Fig. 9. As anticipated, the probability of a Th cell response increases with antigen concentration (thus Q) and the fractional area A. One can use this plot to generate an estimate of c_{thresh}: Q can be estimated for different antigen incubation concentrations from the antigen processing model described earlier, A can be estimated from micrographs, and a plot of P versus QA can be generated from experimental data. A comparison of this plot and Fig. 7 may then allow an estimate to be made. A crude estimate for our experimental system is A = 10 μm^2/500 μm^2 = 0.02 and Q = (4 complexes/μm^2)(500 μm^2) = 2000, so QA = 40. As we find roughly 30–50% of Th cells responding with an increase in $[Ca^{2+}]_i$ after contacting a B cell, Fig. 9 suggests that c_{thresh} is on the order of 30–40 complexes. Note that this is the number of complexes located in the small contact area, presumably greater than the number actually bound by TCR.

A more sophisticated approach to this problem might be made by allowing for not only a probabilistic calculation of the number of MHC-ag complexes present in the initial contact area but also for probabilistic binding kinetics (*e.g.*, 47,48) between MHC-ag and TCR. At present, however, experimental data on the percentage of activated Th cells resulting from single APC/Th cell contacts are too limited to warrant this more detailed approach.

Finally, one is tempted to suggest that the Th cell actively pulls away from the APC and that if the number and strength of TCR/MHC-ag bonds is great enough, the Th cell cannot disengage and the contact area between the two cells

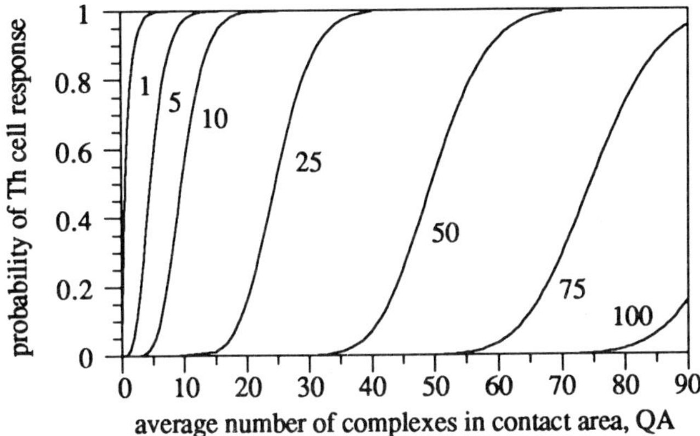

FIGURE 9. Probability of Th cell response as a function of the average number of complexes in the contact area, QA. Each curve represents a different value of the threshold number of complexes that must be "seen" by a Th cell, c_{thresh}.

is triggered to increase. However, recent reports suggest that the low affinity TCR/MHC-ag bonds are unlikely to contribute much to cell-cell adhesion. The more likely players contributing to the strength of interations include CD2/LFA-3 bonds and LFA-1/ICAM-1 bonds (14,49,50).

5.3. Accumulation of TCR/MHC-ag bonds in the contact area. If the initial contact between an APC and Th cell in our experimental system is successful in producing a calcium response in the Th cell, the two cells remain together for the remainder of the experiment, typically 30–90 minutes, and are resistant to fluid shear forces due to pipetting (see Fig. 3). Although most assays of Th cell stimulation do not involve analysis of live single cell conjugates, our results are consistent with observations by Kupfer and Singer (17) of fixed cells and the report by Dustin and Springer (15) that LFA-1 affinity remains elevated for 30–60 min following initial contact between an APC and Th cell.

The purpose of this prolonged contact between an APC and a Th cell is not clear, although it appears necessary for a full Th cell response beyond the initial event of a $[Ca^{2+}]_i$ increase. For example, Goldsmith and Weiss (51) find that sustained calcium increases and 2–4 hours of receptor occupancy may be important to Th cell commitment in the Jurkat cell line. A parallel may easily be drawn to the epidermal growth factor (EGF) receptor system, in which initial receptor/ligand binding is sufficient to generate some signals (*e.g.*, calcium increases, ion fluxes, etc.) but not final commitment.

During this prolonged cell-cell contact, the contact area between the Th cell and APC grows (see Fig. 3). It is likely that a number of molecules redistribute into the contact area and are trapped there by being bound (52-54); presumably MHC-ag complexes will also diffuse into the contact area and be trapped there

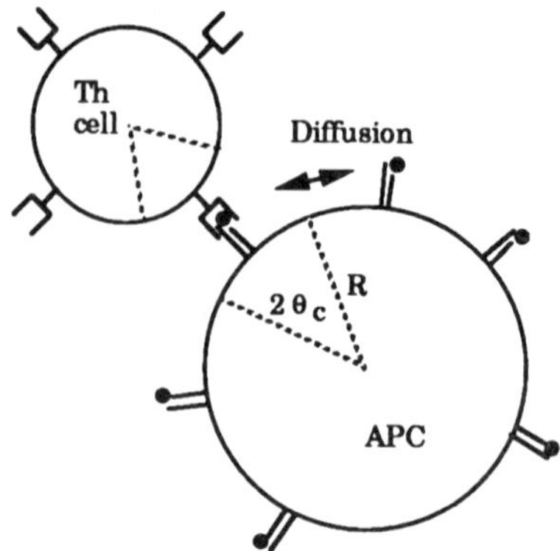

FIGURE 10. Two compartment model for following diffusion of MHC-ag complexes on the APC, diffusion of TCR on the Th cell, and binding of TCR to MHC-ag. Dotted lines denote the distinction between cell surface area inside the contact and cell surface area outside the contact area.

by binding to TCR. If the continued occupancy of TCR is a key factor to the development of a full Th cell response, then it is useful to know the number of occupied TCR as a function of time and MHC-ag concentration (see Fig. 2).

A straightforward estimate of the number of TCR/MHC-ag bonds in the contact area can be obtained by the following approach. We consider a two compartment model for each cell (Fig. 10). The first compartment consists of the cell surface area that is not part of the contact area. In this compartment, cell surface molecules are free to diffuse on the cell surface but cannot bind to their complementary ligand on the other cell. The second compartment consists of the cell surface area that is a part of the contact area. In this region, diffusion as well as binding can occur. Thus we describe both binding and transport (diffusion) events for these cell surface molecules.

We will assume for this simple calculation that the contact area is fixed in size and maintained by receptor/ligand interactions of higher affinity than the TCR/MHC-ag interaction. As mentioned earlier, it is likely, for example, that CD2/LFA-3 bonds and LFA-1/ICAM-1 bonds play a much larger role in adhesion than the TCR/MHC-ag complexes.

Kupfer *et al.* (16), observing cell-cell couples fixed after about 30 minutes of incubation in the same test tube, use antibody staining to show that TCR redistribute from an initially uniform placement on the Th cell and accumulate in the contact area. As the number of TCR on a typical Th cell, on the order of 20,000 molecules, is far greater than the likely number of specific MHC-ag complexes, this accumulation cannot be due solely to trapping of the TCR via MHC-ag binding. For this reason, we assume that TCR can diffuse into the contact area and are then trapped in the contact area by an unspecified mechanism.

Thus at long times, all TCR will be found in the contact area.

Lacking evidence to the contrary, we assume that MHC-ag complexes may diffuse into the contact area and that unbound MHC-ag complexes in the contact area may diffuse out of the contact area. No gross redistribution of MHC-ag molecules has been reported; however, we note that if only the few MHC-ag complexes containing a particular antigen (roughly 100–2000 out of a total of 100,000) were to be trapped in the contact area, such redistribution would not be obvious with current techniques, which utilize antibodies specific only for the MHC molecule and not the MHC-ag complex.

We will assume that there are C_{TOT} (#/cell) MHC-ag complexes on the APC and R_{TOT} (#/cell) TCR on the Th cell. The density of unbound complexes in the contact area is c_{in} (#/μm^2), the density of unbound complexes outside the contact area is c_{out} (#/μm^2), the density of TCR in the contact area is r_{in} (#/μm^2), the density of TCR outside the contact area is r_{out} (#/μm^2), and the density of MHC-ag/TCR bonds in the contact area is b (#/μm^2). MHC-ag complexes can diffuse into the contact area with rate constant $k_{in,c}$ (1/sec). Unbound MHC-ag complexes can diffuse out of the contact area with rate constant k_{out} (1/sec) or bind to free TCR with association rate constant k_1 (μm^2/sec). TCR may diffuse into the contact area with rate constant $k_{in,r}$ (1/sec) but are then trapped in the contact area. MHC-ag/TCR bonds may dissociate with rate constant k_{-1} (1/sec).

The equations describing the evolution of species in the contact area are:

$$\frac{dc_{in}}{dt} = k_{in,c}c_{out}\frac{A_{out,apc}}{A_{in}} - k_{out}c_{in} - k_1 c_{in} r_{in} + k_{-1}b \tag{5.14}$$

$$\frac{dr_{in}}{dt} = k_{in,r}r_{out}\frac{A_{out,tcell}}{A_{in}} - k_1 c_{in} r_{in} + k_{-1}b \tag{5.15}$$

$$\frac{db}{dt} = k_1 c_{in} r_{in} - k_{-1}b \tag{5.16}$$

and are solved together with conservation equations for the total number of MHC-ag complexes and TCR:

$$C_{TOT} = c_{out}A_{out,apc} + (c_{in} + b)A_{in} \tag{5.17}$$

$$R_{TOT} = r_{out}A_{out,tcell} + (r_{in} + b)A_{in} \tag{5.18}$$

A_{in} is the contact area between the two cells, $A_{out,apc}$ is the total area of the APC minus the contact area, and $A_{out,tcell}$ is the total area of the Th cell minus the contact area. The contact area is assumed to remain constant.

5.3.1. *Parameter estimation.* Rate constants for the diffusion of molecules into and out of the contact area can be estimated according to the method of Berg and Purcell (55) and Szabo *et al.* (56). In this method, one first calculates W, the mean time for a molecule at some initial position to reach an absorbing boundary, from the mean capture time equation

$$D_n \nabla_n^2 W + 1 = 0 \tag{5.19}$$

where D_n is the appropriate translational diffusion coefficient and ∇_n^2 is the Laplacian operator in the relevant dimension and coordinate system. Because a finite mean capture time is generated by the initial placement of a molecule in the system and all initial positions are assumed equally likely, there is a constant source term in the equation. The average value of the mean capture time, W_{ave}, is found by averaging W over all possible initial positions of the molecule, and the rate constant k for diffusive motion is the inverse of W_{ave}:

$$k = (W_{ave})^{-1} = \left(\frac{\int_A W dA}{\int_A dA}\right)^{-1} \tag{5.20}$$

For the diffusion of MHC-ag molecules or TCR into the contact area, we assume that molecules travel on a perfect sphere to reach the boundary of the contact area located at $\theta = \theta_c$ (see Fig. 10), and we solve Eqn. (5.19) with the boundary conditions

$$W(\theta_c) = 0 \tag{5.21}$$

$$\left.\frac{dW}{d\theta}\right|_{\theta=0} = 0 \tag{5.22}$$

and then calculate the rate constant from Eqn. (5.20) to obtain:

$$k_{in} = \frac{D}{R^2}\left(\frac{1 + cos\theta_c}{2ln\left(\frac{2}{1-cos\theta_c}\right) - (1 + cos\theta_c)}\right) \tag{5.23}$$

Eqn. (5.23) for k_{in} can be used to calculate both $k_{in,c}$ and $k_{in,r}$ with the use of the appropriate diffusion coefficient D, cell radius R, and critical angle θ_c. One could use Eqns. (5.19,5.20) to estimate a value for k_{out} as well; however, in order to assure that in the absence of binding that MHC-ag complexes would distribute uniformly over the cell, we instead calculate k_{out} from

$$k_{out} = k_{in,c}\left(\frac{A_{out,apc}}{A_{in}}\right) \tag{5.24}$$

We assume for these rate constant calculations that Th cells are perfect spheres with surface areas of 200–500 μm^2, APC are perfect spheres with surface areas of 400–2000 μm^2, and the contact area has a surface area of 10–100 μm^2. The diffusion coefficients for TCR and MHC-ag complexes are assumed to be identical. A typical diffusion coefficient for a cell surface molecule is approximately 10^{-10} cm^2/sec (48,57), although Griffith *et al.* (58) have measured a larger diffusion coefficient of $4x10^{-9}$ cm^2/sec for MHC molecules on TA3 cells.

Matsui *et al.* (59) and Weber *et al.* (60) report a low affinity interaction of TCR with MHC-ag with a K_D on the order of 10^{-6} to 10^{-5} M. However, no kinetic studies that would allow calculation of k_1 or k_{-1} from experimental data have appeared. To estimate a value for k_1, the association rate constant for TCR/MHC-ag binding, we assume that the binding of these two cell surface molecules is diffusion-limited (61, 48 and references therein) and can be calculated from

$$k_1 = \frac{2\pi D}{ln\left(\frac{b}{s}\right)} \tag{5.25}$$

where b is one-half the mean distance between TCR molecules and s is the encounter radius (on the order of 1–10 nm). The appropriate value of the diffusion coefficient D to use is the sum of the diffusivities of the TCR and MHC-ag complex molecules. The estimated value of k_1 can be converted from the units used here, $\mu m^2/sec$, to the more common units of $M^{-1}sec^{-1}$ by multiplying by Avogadro's number and by a thickness of approximately 10 nm (to give an estimate of the local volume of membrane molecules). A value for k_{-1} can be estimated from this value of k_1 and the reported values of $K_D = k_{-1}/k_1$.

5.3.2. *Model predictions.* Eqns. (5.14–5.18) were solved numerically. Predictions of this simple model for diffusion and binding within the contact area between two cells are shown in Fig. 11. The fraction f of MHC-ag complexes bound by TCR ($f = bA_{in}/C_{TOT}$) is plotted. For the same diffusion coefficient and K_D (10^{-6} M) of the TCR/MHC-ag interaction, curves a-c represent several different scenarios in terms of cell sizes and contact areas. In curve a, the worst case scenario of a small contact area and large APC and Th cells is shown. In curve b, the best case scenario of a large contact area and small APC and Th cells is shown. In curve c, a more typical scenario is used; the contact area is 50 μm^2 and the APC and Th cells have total surface areas of 400 and 250 μm^2, respectively.

A comparison of curves c and d in Fig. 11 allows the effect of affinity to be seen. Raising the value of k_{-1} (calculated to correspond to a value for K_D of 10^{-5} M) decreases the number of MHC-ag complexes bound at any time. Finally, a comparison of curves c and e allows the effect of the diffusion coefficient D to be seen. Note that a change in D also changes the values of k_1 (and thus k_{-1})

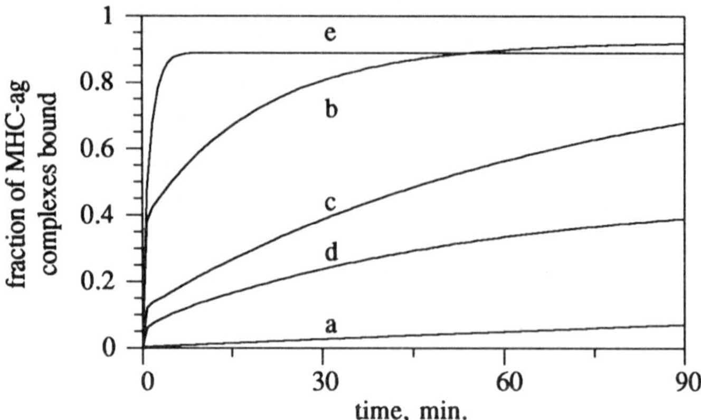

FIGURE 11. Fraction of MHC-ag complexes bound by TCR as a function of time. For all curves, R_{TOT} = 20,000. Curve a: C_{TOT} = 200, A_{in} = 10 μm^2, $A_{out,apc}$ = 2000 μm^2, k_1 = 0.05 μm^2/sec, k_{-1} = 0.3/sec, D=0.01 μm^2/sec, and $A_{out,tcell}$ = 500 μm^2. Curve b: C_{TOT} = 2000, A_{in} = 100 μm^2, $A_{out,apc}$ = 150 μm^2, k_1 = 0.05 μm^2/sec, k_{-1} = 0.3/sec, D=0.01μm^2/sec, and $A_{out,tcell}$ = 100 μm^2. Curve c: C_{TOT} = 800, A_{in} = 50 μm^2, $A_{out,apc}$ = 350 μm^2, k_1 = 0.05 μm^2/sec, k_{-1} = 0.3/sec, D=0.01μm^2/sec, and $A_{out,tcell}$= 200 μm^2. Curve d: Same parameters as curve c except that k_{-1} = 3.0/sec. Curve e: Same parameters as curve c except that k_1 = 2.5 μm^2/sec, k_{-1} = 15.0/sec, and D = 0.5 μm^2/sec. The steady state fractions f^{ss} of MHC-ag complexes bound by TCR are: curve a, f^{ss} = 0.62; curve b, f^{ss} =0.92; curve c, f^{ss} =0.89; curve d, f^{ss} =0.45; and curve e, f^{ss} =0.89

through Eqn. (5.25). As expected, an increase in D significantly increases the number of MHC-ag complexes bound at early times.

One can calculate that the steady state solution to Eqns. (5.14–5.18) is:

$$c_{in}^{ss} = \frac{-\alpha + \sqrt{\alpha^2 + \left(\frac{4k_1 k_{in,c} C_{TOT}}{k_{-1} A_{in}}\right)(k_{in,c} + k_{out})}}{2\frac{k_1}{k_{-1}}(k_{in,c} + k_{out})} \tag{5.26}$$

$$r_{in}^{ss} = \frac{R_{TOT}}{A_{in}\left(1 + \frac{k_1}{k_{-1}} c_{in}^{ss}\right)} \tag{5.27}$$

$$b^{ss} = \frac{k_1}{k_{-1}} c_{in}^{ss} r_{in}^{ss} \tag{5.28}$$

where

$$\alpha = \frac{k_1 k_{in,c}}{k_{-1} A_{in}} (R_{TOT} - C_{TOT}) + k_{in,c} + k_{out} \tag{5.29}$$

For the curves shown in Fig. 11, the steady state values of the fraction f of MHC-ag found in TCR/MHC-ag complexes are listed in the figure legend. Note that for all but curve a, which represents a worst case scenario in terms of cell sizes and diffusivity, the steady state value of f ($f^{ss} = b^{ss} A_{in}/C_{TOT}$) gives a reasonable estimate for the number of TCR/MHC-ag bonds after 1–2 hours of cell-cell contact. Allowing for a linear increase in the contact area from a small initial value on the order of 10 μm^2 to a value at about 30 min of 50–100 μm^2 does not significantly alter this conclusion (calculations not shown). If the binding of MHC-ag complexes to TCR can be considered to be nearly at steady state for much of the time the APC and Th cells are together, later analysis of the events of Th cell signal transduction—for example, secretion of IL-2 and expression of IL-2 receptors—might be related to the steady state level of MHC-ag/TCR bonds.

Finally, we return briefly to the observation of Kupfer *et al.* (16) mentioned earlier in this section. Using a qualitative fluorescent antibody technique, these investigators found that after approximately 30 min of cell-cell contact, the majority of the TCR are found in the contact area. The simplest possibility, that consistent with our equations above, is that diffusion and trapping of TCR within the contact area is sufficient to account for this observation. Alternatively, one might postulate that some type of membrane current or active dragging of TCR by the cytoskeleton is needed for such rapid redistribution of TCR. For curves a–e shown in Fig. 11, we can use Eqns. (5.14–5.18) to calculate that at 30 min, 15%, 95%, 66%, 66%, and 100%, respectively, of the TCR are found in the contact area. Thus for all but curve a, the worst case scenerio, simple diffusion and trapping of TCR is sufficient to account for the observations of Kupfer and co-workers.

6. Discussion

As described previously (Fig. 1), the process by which Th cells become activated is complex, for it involves a step distinct from the Th cell (preparation of MHC-ag complexes by APC), a step involving both the APC and Th cell (cell adhesion/signal transduction), and steps involving only the Th cell (*e.g.*, IL-2 secretion and IL-2 receptor expression). In this work, we have attempted to quantitate some of the underlying relationships between key variables in the steps of this process. In order to develop clinical treatments or methods through which the Th cell response could be enhanced or inhibited, we feel that such quantitation could be extremely useful.

The first model we describe enables us to calculate the number of MHC-ag complexes expressed by an APC as a function of time, antigen concentration, method of antigen internalization, competing protein concentration, and key cellular parameters. A quantitative understanding of even this first step of antigen presentation may guide manipulation of the immune response. For example, Adorini *et al.* (42) suggest that non-immunogenic molecules with an affinity for the MHC class II molecule might be used to block the response to an antigen by reducing the number of MHC-ag complexes available for binding to TCR, and state the possibility of "immunomanipulation at the level of antigen presentation" as a treatment for autoimmune disease. Many others now favor the idea of "designer peptides" which would compete with naturally present peptides for MHC binding (62,63). The degree of competition of these designer peptides with immunogenic peptides can be influenced, for example, by targeting the designer peptides to rapidly internalizing receptors and by manipulating association and dissociation rate constants (*e.g.*, 64–66). The influence of receptor internalization rates and association and dissociation rate constants is easily calculated with our model. Current difficulties in accurately measuring the number of MHC-ag complexes on a cell make the usefulness of such a model even more apparent.

We also present the hypothesis that the Th cell must "see" at least a threshold number of MHC-ag complexes during an initial limited contact between the Th cell and the APC. This suggests that for many antigen concentrations (and corresponding numbers of MHC-ag complexes presented by the APC), the responses of individual Th cells may be heterogeneous. In the most exaggerated case, the response of a Th cell could be all-or-none. It is not yet clear whether Th cells respond in an all-or-none or graded fashion to the presence of an APC bearing compatible MHC-ag complexes, although our preliminary data on increases in intracellular free calcium in Th cells are consistent with an all-or-none response at the level of calcium. If the threshold number of MHC-ag complexes needed to trigger a Th cell response is very small, it may be difficult to completely eliminate a Th cell response by using designer peptides to compete with an immunogenic peptide. In this case, it might be more effective to block MHC or TCR molecules themselves; Wraith and co-workers (62) note that anti-MHC therapy holds promise for diminishing an autoimmune response and Smith and Allen (67) discuss anti-TCR approaches.

Third, we describe a model to follow the diffusion and binding of MHC-ag complexes and TCR. This model is intended to apply to cases in which a Th cell has indeed "seen" sufficient MHC-ag complexes to initiate a full response. In this simple model, we ignore the details of contact area formation and the participation of other complementary cell surface molecules in the development of cell-cell adhesion. In fact, we assume that the affinity of MHC-ag/TCR complexes is immaterial to the contact area and that these molecules are likely to play a major role in signaling rather than adhesion. Most interestingly, we find that for reasonable parameter values the binding of MHC-ag complexes to TCR can be considered to be nearly at steady state for much of the time the APC and Th cells are together. This suggests that events in the Th cell response which occur hours after APC/Th cell contact—for example, IL-2 secretion and IL-2 receptor expression—might be able to be related to the steady state level of MHC-ag/TCR bonds. A similar approach has been found useful in the epidermal growth factor (EGF) system, for which the cell proliferation rate was found to be linearly related to the number of occupied receptors at steady state (68,69). Mathematically, this would be a useful simplification of a complicated dynamic problem. On the other hand, many reports now suggest that other bound Th cell surface molecules can also play a signaling role (26). The calculation of the number of these molecules occupied with time as a function of key cellular parameters may be important, and more sophisticated models of cell-cell adhesion than that presented here may be needed for this system (70-72).

Finally, we note that ultimately it will be important to incorporate what is learned from experiments and models at the single cell level into whole body models of immune system function. Such models should take into account the availability of different peptides at different locations, the relative roles and locations of different types of APC, and the environment in which APC and Th cells actually make contact.

REFERENCES

1. Sinha, A.A., M.T. Lopez, and H.O. McDevitt, Autoimmune diseases: The failure of self tolerance. Science **248** (1990), 1380-1388.

2. Chestnut, R.W., and H.M. Grey, Antigen-presenting cells and mechanisms of antigen presentation, CRC Crit. Rev. Immunol. **5** (1985), 263-316.

3. Unanue, E.R., and P.M. Allen, The basis for the immunoregulatory role of macrophages and other accessory cells, Science **236** (1987), 551-557.

4. Brodsky, F.M. and L.E. Guagliardi, The cell biology of antigen processing and presentation, Annu. Rev. Immunol. **9** (1991), 707-744.

5. Babbitt, B.P., P.M. Allen, G. Matsueda, E. Haber, and E.R. Unanue, Binding of immunogenic peptides to Ia histocompatibility molecules, Nature (Lond.) **317** (1985), 359-361.

6. Buus, S., A. Sette, and H.M. Grey, The interaction between protein-derived immunogenic peptides and Ia, Immunol. Rev. **98** (1987), 115-141.

7. Guillet, J., M. Lai, T.J. Briner, S. Buus, A. Sette, H.M. Grey, J.A. Smith, and M.L. Gefte, Immunological self, nonself discrimination, Science **235** (1987), 865-870.

8. Ziegler, H.K., and E.R. Unanue, Decrease in macrophage antigen catabolism caused by ammonia and chloroquine is associated with inhibition of antigen presentation to T cells, Proc. Natl. Acad. Sci. USA **79**(1981), 175-178.

9. Lakey, E.K., L.A. Casten, W.L. Niebling, E. Margoliash, and S.K. Pierce, Time dependence of B cell processing and presentation of peptide and native protein antigens, J. Immunol. **140** (1988): 3309-3314.

10. Casten, L.A. and S.K. Pierce, Receptor-mediated B cell antigen processing: Increased antigenicity of a globular protein covalently coupled to antibodies specific for B cell surface structures, J. Immunol. **140** (1988), 404-410.

11. Allen, P.M., B.P. Babbitt, and E.R. Unanue, T-cell recognition of lysozyme: The biochemical basis of presentation, Immunol. Rev. **98** (1987), 170-187.

12. Watts, T.H., and H.M. McConnell, Biophysical aspects of antigen recognition by T cells, Annu. Rev. Immunol. **5** (1987), 461-475.

13. Bierer, B.E., B.P. Sleckman, S.E. Ratnofsky, and S.J. Burakoff, The biologic roles of

CD2, CD4, and CD8 in T-cell activation. Ann. Rev. Immunol. **7** (1989), 579-599.

14. Springer, T.A., Adhesion receptors of the immune system, Nature (Lond.) **346** (1990), 425-434.

15. Dustin, M.L., and T.A. Springer, T-cell receptor cross-linking transiently stimulates adhesiveness through LFA-1, Nature (Lond.) **341** (1989), 619-624.

16. Kupfer, A., S.J. Singer, C.A. Janeway, and S.L. Swain, Coclustering of CD4 (L3T4) molecule with the T-cell receptor is induced by specific direct interaction of helper T cells and antigen-presenting cell, Proc. Natl. Acad. Sci. USA **84** (1987), 5888-5892.

17. Kupfer, A., and S.J. Singer, The specific interaction of helper T cells and antigen-presenting B cells. IV. Membrane and cytoskeletal reorganizations in the bound T cell as a function of antigen dose, J. Exp. Med. **170** (1989), 1697-1713.

18. Fox, B.S., H. Quill, L. Carlson, and R.H. Schwartz, Quantitative analysis of the T cell response to antigen and planar membranes containing purified Ia molecules, J. Immunol. **138** (1987), 3367-3374.

19. Roosnek, E., S. Demotx, G. Corradin, and A. Lanzavecchia, Kinetics of MHC-antigen complex formation on antigen-presenting cells, J. Immunol. **140** (1988), 4079-4082.

20. Gardner, P., Calcium and T lymphocyte activation, Cell **59** (1989), 15-20.

21. Donnadieu, E., D. Cefai, Y.P. Tan, G. Paresys, G. Bismuth, and A. Trautmann, Imaging early steps of human T cell activation by antigen-presenting cells, J. Immunol. **148** (1992), 2643-2653.

22. Putney, J.W., H. Takemura, A.R. Hughes, D.A. Horstman, and O. Thastrup, How do inositol phosphates regulate calcium signaling?, FASEB J. **3** (1989), 1899-1905.

23. Berridge, M.J, Inositol trisphosphate and calcium signalling, Nature **361** (1993), 315-325.

24. Tsien, R.Y., T.J. Rink, and M. Poenie, Measurement of cytosolic free Ca^{2+} in individual small cells using fluorescence microscopy with dual excitation wavelengths, Cell Calcium **6** (1985), 145-157.

25. Tsien, R.Y., Fluorescent indicators of ion concentrations, Methods Cell Biol. **30** (1989), 127-156.

26. Altman, A., K.M. Coggeshall, and T. Mustelin, Molecular events mediating T cell activation, Adv. Immunol. **48** (1990), 227-360.

27. Geppert, T.D., L.S. Davis, H. Gur, M.C. Wacholtz, and P.E. Lipsky, Accessory cell signals involved in T-cell activation, Immunol. Rev. **117** (1990), 5-66.

28. Klausner, R.D., and L.E. Samelson, T cell antigen receptor activation pathways: The tyrosine kinase connection, Cell **64** (1991), 875-878.

29. Esch, T.R., Structure and function within the T-cell antigen recognition complex, Ph.D. Thesis, Dept. of Microbiology and Immunology, Univ. of Michigan (1989).

30. Grynkiewicz, G., M. Poenie, R.Y. Tsien, A new generation of Ca^{2+} indicators with greatly improved fluorescence properties, J. Biol. Chem. **260** (1985), 3440-3450.

31. Harding, C.V., R.W. Roof, and E.R. Unanue, Turnover of Ia-peptide complexes is facilitated in viable antigen-presenting cells: Biosynthetic turnover of Ia vs. peptide exchange, Proc. Natl. Acad. Sci. USA **86** (1989), 4230-4234.

32. Lanzavecchia, A., S. Siervo, and D. Scheidegger, On the role of surface Ig in antigen presentation to T cells, *Processing and Presentation of Antigens*, Academic Press, Inc., Orlando, FL, 1988.

33. Demotz, S., H.M. Grey, and A. Sette, The minimal number of class II MHC-antigen complexes needed for T-cell activation, Science **249** (1990), 1028-1030.

34. Harding, C.V., and E.R. Unanue, Quantitation of antigen-presenting cell MHC class II/peptide complexes necessary for T-cell stimulation, Nature (Lond.) **346** (1990), 574-576.

35. Singer, D.F., and J.J. Linderman, The relationship between antigen concentration, antigen internalization, and antigenic complexes: Modelling insights into antigen processing and presentation, J. Cell Biol. **111** (1990), 55-68.

36. Harding, C.V., D.S. Collins, J.W. Slot, H.J. Geuze, and E.R. Unanue, Liposome-encapsulated antigens are processed in lysosomes, recycled, and presented to T cells, Cell **64** (1991), 303-401.

37. Rothbard, J.B. and M.L. Gefter, Interactions between immunogenic peptides and MHC proteins, Annu. Rev. Immunol. **9** (1991), 527-565.

38. Marsh, E.W., D.P. Dalke, and S.K. Pierce, Biochemical evidence for the rapid assembly and disassembly of processed antigen-major histocompatibility complex class II complexes in acidic vesicles of B cells, J. Exp. Med. **175** (1992), 425-436.

39. Chu, C.T. and S.V. Pizzo, Receptor-mediated antigen delivery into macrophages: Complexing antigen to α_2-macroglobulin enhances presentation to T cells, J. Immunol. **150** (1993), 48-58.

40. Tse, D.B., C.R. Cantor, J. McDowell, and B. Pernis, Recycling class I MHC antigens: dynamics of internalization, acidification, and ligand-degradation in murine T- lymphoblasts, J. Mol. Cell. Immunol. **2** (1986), 315-329.

41. Lanzavecchia, A., P.A. Reid, and C. Watts, Irreversible association of peptides with class II MHC molecules in living cells, Nature **357** (1992), 249-252.

42. Adorini, L. S. Muller, F. Cardinauz, P.V. Lehmann, F. Falcioni, and Z.A. Nagy, *In vivo* competition between self peptides and foreign antigens in T-cell activation, Nature (Lond.) **334** (1988), 623-625.

43. Lorenz, R.G., and P.M. Allen, Direct evidence for functional self-protein/Ia-molecule complexes in vivo, Proc. Natl. Acad. Sci. USA **85** (1988), 5220-5223.

44. Adorini, L., E. Appella, G. Doria, F. Cardinaux, and Z.A. Nagy, Competition for antigen presentation in living cells involves exchange of peptides bound by class II MHC molecules, Nature (Lond.) **342** (1989), 800-803.

45. Lorenz, R.G., J.S. Blum, and P.M. Allen, Constitutive competition by self proteins for antigen presentation can be overcome by receptor-enhanced uptake, J. Immunol. **144** (1990), 1600-1606.

46. Singer, D.F., and J.J. Linderman, Antigen processing and presentation: How can foreign antigen be recognized in a sea of self proteins?, J. Theor. Biol. **151** (1991), 385-404.

47. Cozens-Roberts, C., D.A. Lauffenburger, and J.A. Quinn, Receptor-mediated cell attachment and detachment kinetics: I. Probabilistic model and analysis, Biophys. J. **58** (1990), 841-856.

48. Lauffenburger, D.A., and J.J. Linderman, *Receptors: Models for Binding, Trafficking, and Signaling*, Oxford University Press, New York, in press 1993.

49. Sung, K.-L. P., P. Kuhlman, F. Maldonado, B.A. Lollo, S. Chien, and A.A. Brian, Force contribution of the LFA-1/ICAM-1 complex to T cell adhesion, J. Cell Sci. **103** (1992), 259-266.

50. Williams, A.F., and A.D. Beyers, At grips with interactions, Nature **356** (1992), 746-747.

51. Goldsmith, M.A., and A. Weiss, Early signal transduction by the antigen receptor without commitment to T cell activation, Science **240** (1988), 1029-1031.

52. McCloskey, M.A. and M. Poo, Contact-induced redistribution of specific membrane components: Local accumulation and development of adhesion, J. Cell Biol. **102** (1986), 2185-2196.

53. Chan, P., M.B. Lawrence, M.L. Dustin, L.M. Ferguson, D.E. Golan, and T.A. Springer, Influence of receptor lateral mobility on adhesion strenghening between membranes containing LFA-3 and CD2, J. Cell Biol. **115** (1991), 245-255.

54. Singer, S.J, Intercellular communication and cell-cell adhesion, Science **255** (1992), 1671-1677.

55. Berg, H.C., and E.M. Purcell, Physics of chemoreception, Biophys. J. **20** (1977), 193-219.

56. Szabo, A., K. Schulten, and Z. Schulten, First passage time approach to diffusion controlled reactions, J. Chem. Phys. **72** (1980), 4350-4357.

57. Gennis, R.B., *Biomembranes: Molecular Structure and Function*, Springer-Verlag, New York, 1989.

58. Griffith, I.J., A. Ghogawala, N. Nabavi, D.E. Golan, A. Myer, D.J. McKean, and L.H. Glimcher, Cytoplasmic domain affects membrane expression and function of an Ia molecule, Proc. Natl. Acad. Sci. USA **85** (1988), 4847-4851.

59. Matsui, K. J., J.J. Boniface, P.A. Reay, H. Schild, B. Fazekas de St. Groth, and M.M. Davis, Low affinity interacton of peptide-MHC complexes with T cell receptors, Science **254** (1991), 1788-1791.

60. Weber, S., A. Traunecker, F. Oliveri, W. Gerhard, and K. Karjalainen, Specific low-affinity recognition of major histocompatibility complex plus peptide by soluble T-cell receptor, Nature **356** (1992), 793-796.

61. Lauffenburger, D., and C. DeLisi, Cell surface receptors: Physical chemistry and cellular regulation, Internatl. Rev. Cytology **84** (1983), 269-301.

62. Wraith, D.C., H.O. McDevitt, L. Steinman, and H. Acha-Orbea, T cell recognition as the target for immune intervention in autoimmune disease, Cell **57** (1989), 709-715.

63. Alexander, J., K. Snoke. J. Ruppert, J. Sidney, M. Wall, S. Southwood, C. Oseroff, T. Arrhenius, F.C.A. Gaeta, S.M. Colon, H.M. Grey, and A. Sette, Functional consequences of engagement of the T cell receptor by low affinity ligands, J. Immunol. **150** (1993), 1-7.

64. Snider, D.P., and D.M. Segal, Targeted antigen presentation using crosslinked antibody heteroaggregates, J. Immunol. **139** (1987), 1609-1616.

65. Snider, D.P., and D.M. Segal, Efficiency of antigen presentation after antigen targeting to surface IgD, IgM, MHC, FcgRII, and B220 molecules on murine splenic B cells, J. Immunol. **143** (1989), 56-65.

66. Sette, A., J. Sidney, M. Albertson, C. Miles, S.M. Colon, T. Pedrazzini, A.G. Lamont, and H.M. Grey, A novel approach to the generation of high affinity class II-binding peptides, J. Immunol. **145** (1990), 1809-1813.

67. Smith, S.C., and P.M. Allen, The recognition of self-antigens and autoimmune disease, Curr. Opinion in Immunol. **3** (1991), 22-25.

68. Knauer, D.J., H.S. Wiley, and D.D. Cunningham, Relationship between epidermal growth factor receptor occupancy and mitogenic response: quantitative analysis using a steady-state model system, J. Biol. Chem. **259** (1984), 5623-5631.

69. Starbuck, C. and D.A. Lauffenburger, Mathematical model for the effects of epidermal growth factor receptor trafficking dynamics on fibroblast proliferation responses, Biotech. Prog. **8** (1992), 132-143.

70. Dembo, M., D.C. Torney, K. Saxman, and D. Hammer, The reaction-limited kinetics of membrane-to-surface adhesion and detachment, Proc. R. Soc. Lond. B **234** (1988), 55-83.

71. Evans, E., Mechanics of cell deformation and cell-surface adhesion, *Physical Basis of Cell-Cell Adhesion* (P. Bongrand, ed.), CRC Press, Boca Raton, FL, 1988.

72. Hammer, D.A., and D.A. Lauffenburger, A dynamical model for receptor-mediated cell adhesion to surfaces, Biophys. J. **52** (1987), 475-487.

DEPARTMENT OF CHEMICAL ENGINEERING, UNIVERSITY OF MICHIGAN, ANN ARBOR, MICHIGAN 48109

E-mail address: jennifer.linderman@um.cc.umich.edu

Lectures on Mathematics in the Life Sciences
Volume **24**, 1994

Aggregation of Cell Surface Receptors

BYRON GOLDSTEIN and CARLA WOFSY

ABSTRACT. The aggregation of cell surface receptors, induced by the binding of hormones, interleukins, antigens, and other extracellular signaling molecules, is essential for triggering diverse cellular responses. Ligands with two or more binding sites can induce receptor aggregation by binding simultaneously to two or more receptors, crosslinking them. We present a model for estimating equilibrium constants characterizing the crosslinking of cell surface receptors. Using the model, we show how, and under what conditions, crosslinking can explain high affinity binding observed in systems where isolated receptors bind the ligand with relatively low affinity. The model can also be exploited to estimate, from binding data, the degree to which steric hindrance limits crosslinking.

When receptors and ligands both have multiple binding sites, large crosslinked aggregates can form. In the simplest example, the cell surface receptor is a bivalent antibody (i.e., an antibody with two binding sites), the ligand (antigen) is also bivalent, and the only aggregates that can form are chains or closed rings. Even in this case, following the kinetics of aggregate formation is not feasible without some approximation to simplify the system of differential equations for concentrations of all possible aggregates. We review the equivalent site approximation, which simplifies the kinetic analysis substantially, and discuss conditions (and alternative approaches in special cases) when the approximation breaks down.

1. Introduction

Cells must constantly sense their environment (ions, hormones, growth factors, etc.) and respond to it. The macromolecules that cells detect, and the

---

1991 Mathematics Subject Classification, Primary: 92C05.
Supported by NIH Grant GM355556 and NSF Grant DMS9101969.
This paper is in final form and no version of it will be submitted for publication elsewhere.
The work for this article was performed under the auspices of the U.S. Department of Energy.

concentrations at which they detect them, are determined by the cell surface receptors they express. The set of receptors characteristic of a particular cell type is determined by the special functions the cell performs and the sustenance the cell requires to perform them. At any moment, the subset of receptors a cell expresses, and the levels of expression, are determined by the history of the cell, e.g., whether it has been exposed to a particular growth factor, whether it has been in a high serum cholesterol environment, whether it has been in the presence of a chemoattractant. The concentrations and types of receptors appearing on the cell surface change with time, some rapidly, some slowly, up and down regulating in response to their changing environment.

Once a particular macromolecule binds to a receptor, this information must be transduced across the cell membrane if the cell is to respond. If the information transfer is successful, the binding of the ligand to the receptor will initiate a series of events culminating in the activation or suppression of one or more cellular responses. For receptors that span the membrane, the possibility arises that information can be passed directly from their external domains where the ligands bind, along their membrane spanning regions to their cytoplasmic domains. However, for many receptors information is either not transmitted across the plasma membrane in this way, or this is not the sole way information is passed. Many types of receptors aggregate with additional membrane proteins or among themselves to initiate a cell signal.

One can divide ligands that induce receptor aggregation into two broad groups, those that are multivalent and aggregate receptors by physically bridging two or more receptors, and those that are monovalent and aggregate receptors by inducing a change in the receptor that allows aggregation. We shall look at both types of ligand-induced receptor aggregation and try to see what questions arise when one tries to model these processes. We begin by looking at ligands that participate only in the formation of small aggregates. These include many of the growth factors and cytokines. We then consider large aggregate formation, such as occurs when an antigen interacts with cell surface antibody.

2. Cytokines and the Formation of Small Aggregates

2.1. A Simple Example: The Formation of Receptor Dimers. It is instructive to start with the simplest example of receptor aggregation involving a multivalent ligand: the binding of a symmetric bivalent ligand to a monovalent receptor, where in the absence of the ligand the receptor does not aggregate. In this system the only aggregate that can form consists of two receptors bridged by a single ligand: a receptor dimer. We start by considering equilibrium binding. Figure 1 summarizes the reactions that can occur. The ligand binds to the first receptor with single site equilibrium constant K and to the second receptor with single site equilibrium crosslinking constant K_X. We assume that the system equilibrates rapidly so that for the times of interest we can neglect the loss of receptors due to internalization or shedding and the gain of receptors due to recycling and insertion of newly synthesized receptors into the plasma membrane. Under these conditions, R_T, the total number of receptors on the cell surface, is conserved. If R is the concentration of free receptors and C the concentration of free ligands, then from the law of mass action D, the total concentration of receptors in dimers, and R_B, the total concentration of bound receptors, are

(1) $$D = 2KK_XCR^2$$

(2) $$R_B = 2KCR + 2KK_XCR^2 \ .$$

The equation for the conservation of receptors,

(3) $$R_T = R + 2KCR + 2KK_XCR^2$$

is quadratic and can be solved for R. Thus, if the equilibrium constants and the total receptor concentration are known, one can, for example, calculate the concentration of receptors in dimers as a function of the free ligand concentration. The total ligand concentration, C_T, is also conserved,

(4) $$C_T = C + 2KCR + KK_XCR^2 \ .$$

In many experimental situations the amount of ligand bound to the surfaces of cells is negligible compared to the total ligand concentration and one can take $C \approx C_T$. If one can't make this approximation, then one must solve Eqs. (3) and (4) numerically for C and R.

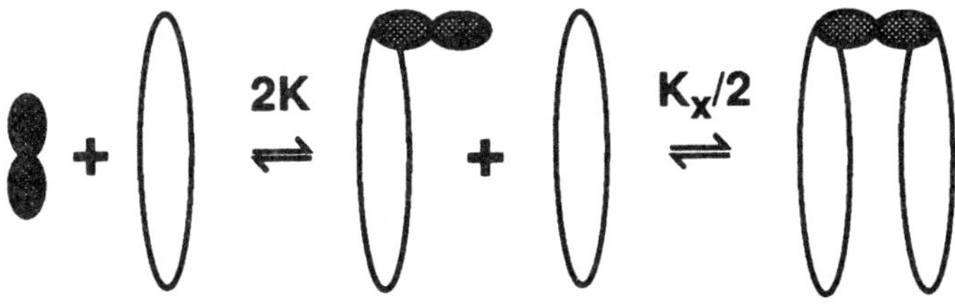

Figure 1. Equilibrium binding of a symmetric bivalent ligand to a monovalent receptor.

It is useful to introduce the following nondimensional receptor and ligand concentrations: $r = R/R_T$ and $c = 2KC$, and nondimensional equilibrium crosslinking constant: $k_x = K_XR_T$. In terms of these quantities the conservation law becomes

(5) $$1 = (1 + c)r + ck_xr^2$$

with positive solution

(6) $$r = (1 + c)(-1 + \sqrt{1 + \delta})/(2k_xc)$$

where

$$\delta = \frac{4k_x c}{(1+c)^2} \quad . \tag{7}$$

The fraction of receptors in dimers is

$$d = ck_x r^2 = 1 - (1+c)r = (\delta + 2 - 2\sqrt{1+\delta})/\delta \tag{8}$$

and for $\delta \ll 1$

$$d \approx \delta/4 \tag{9}$$

The dimer concentration, D, has some interesting properties. Because D depends on C only through the variable δ, one can show that a plot of D versus $\log(C)$ is a symmetric bell shaped curve that has a maximum when the free ligand concentration equals C_{max}, where

$$C_{max} = \frac{1}{2K} \quad . \tag{10}$$

In addition, one can show that at $C = C_{max}$, ligand binding is half maximal with $R_T/2$ ligands bound. Therefore the total molar ligand concentration at which dimer formation is maximal is

$$C_{Tmax} = \frac{1}{2K} + \frac{R_T}{2} \tag{11}$$

where $R_T = \bar{R}\rho/6.02 \times 10^{20}$ and $\bar{R}$ is the total number of receptors per cell, ρ is the number of cells/ml and K is in M^{-1}. With these results in mind, we now look at some data from a receptor system where the ligand is thought to be bivalent and the receptor monovalent.

The first receptor we consider, the platelet-derived growth factor (PDGF) receptor, is one of a family of receptors that binds specific growth factors. As the name implies, cells that display receptors for a particular growth factor can be stimulated by that growth factor to divide and differentiate. Many growth factor receptors have similar primary structures consisting of one protein chain with an extracellular ligand-binding domain, a single transmembrane domain and a cytoplasmic tyrosine kinase domain (reviewed in Yarden and Ullrich, 1988; Ullrich and Schlessinger, 1990). At physiological growth factor concentrations, binding of the growth factor induces receptor dimerization and rapid phosphorylation of multiple tyrosine residues along the growth factor receptor's cytoplasmic domain. These newly created phosphotyrosine residues act as binding sites for specific cytoplasmic signaling molecules, places where signaling molecules can dock, reside while undergoing additional interactions with the receptor and possibly other membrane and cytoplasmic proteins, and then move off (reviewed in Pawson and Gish, 1992). Receptor dimerization appears to be crucial in this activation pathway. In the absence of growth factors, anti-receptor antibodies (but not their monovalent Fab fragments) that can bridge two receptors induce tyrosine phosphorylation (Rönnstrand et al.,

1988; Spaargaren et al., 1991). Dimerization juxtaposes cytoplasmic domains that phosphorylate each other, i.e., the tyrosine kinase domain in the cytoplasmic region of one receptor phosphorylates tyrosines along the cytoplasmic domain of its dimer partner (Honegger et al., 1989; Ohtsuka et al., 1990; Kelly et al., 1991; Bellot et al., 1991). In addition, the dimer is an activated kinase and can phosphorylate tyrosines on other proteins (Heldin et al., 1989).

Platelet-derived growth factor (PDGF) is a disulfide-bonded dimeric protein composed of two chains. The chains can be of two types, A or B, that are highly homologous, but not identical. PDGF has been found in all three isoforms, PDGF-AA, PDGF-BB, and PDGF-AB. There are also two receptors for PDGF, an α and β type (for reviews see Hart and Bowen-Pope, 1990; Williams, 1989). Both PDGF-BB and PDGF-AB aggregate β type PDGF receptors, but PDGF-AA does not (Heldin et al., 1989).

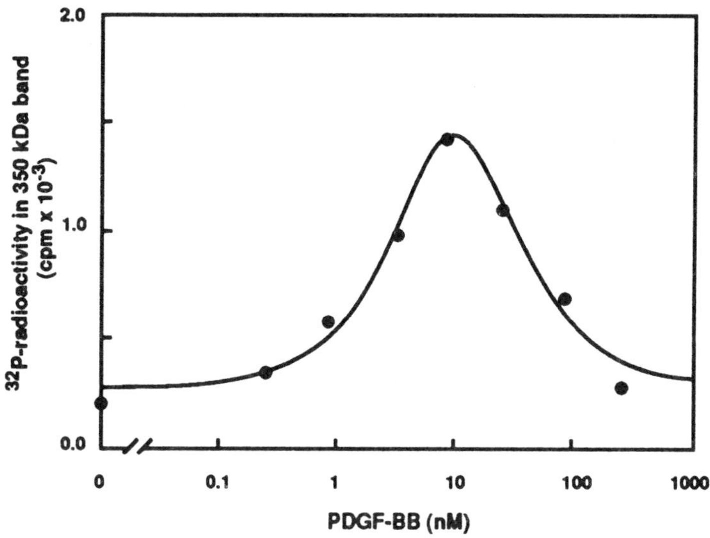

Figure 2. Phosphorylation of receptor dimers as a function of PDGF concentration (•, data from Heldin et al., 1989), fit (—) by a model for dimer formation under the assumption that phosphorylation is a linear function of the fraction of receptors in dimers.

There are few published measurements of receptor dimer concentrations as functions of their ligand concentrations. For the PDGF receptor there are experiments, such as that shown in Figure 2, where the level of phosphorylation in receptor dimers is measured as a function of the PDGF concentration (Heldin et al., 1989). Since after PDGF receptor dimers form, receptor pairs rapidly phosphorylate tyrosines on their partners' cytoplasmic domains, one expects the rise in phosphorylation to be a measure of the rise in dimer concentration. A plot of phosphorylation in PDGF receptor dimers versus PDGF concentration should exhibit the same symmetry properties and location of the maximum as the dimer concentration. Thus, in the experiment in Figure 2 where PDGF-BB was used to induce dimerization of purified β type PDGF receptors, Eq. (11) should hold and the curve should have a maximum when the total PDGF concentration equals $1/(2K) + R_T/2$, where K is the single site equilibrium constant for binding of a B site on PDGF-BB to a β type PDGF receptor and R_T is the total concentration of receptors. The solid curve in Figure 2 is a theoretical fit to the data. We assumed that the level of phosphorylation of the receptors is

$$P = P_0 + (P_D - P_0)d \tag{12}$$

where P_0 is the level of PDGF receptor phosphorylation in the absence of PDGF, P_D is the level of phosphorylation when all receptors are in dimers, and d is calculated from Eq. (8). From the fit we find that $K = (2 \pm 1) \times 10^8$ M^{-1}. The binding of PDGF-BB to β type receptors on cell surfaces is usually reported to be of high affinity, about ten times higher than the value for K determined in Figure 2 (Severinsson et al., 1989; Kelley et al., 1991). It has long been known that bivalent antibodies often bind to cell surface epitopes with much higher apparent affinities than their monovalent Fab fragments. In the next section we review how bivalency can lead to this effect.

2.2. What is Meant by High Affinity Binding to Cell Surface Receptors? Many of the growth factors and other cytokines appear to be bound by their cell surface receptors with high affinity, i.e., with apparent equilibrium constants of order $10^{10} - 10^{11}$ M^{-1}. Cells are able to bind and respond to these ligands at picomolar concentrations.

The conclusion that the binding of growth factors to their cell surface receptors is high affinity comes of course from binding studies. The binding studies are interpreted through the use of Scatchard plots (Scatchard, 1949). Recall that a Scatchard plot is a plot of the equilibrium ratio of bound to free ligand concentrations, B/C, as a function of the bound ligand concentration, B. If both the ligand and the receptor are monovalent and homogeneous and the receptors bind independently, then $B = KC(R_T - B)$; dividing by C, one obtains a straight-line Scatchard plot with slope $-K$ and B intercept R_T.

Often Scatchard plots for cytokines are curved, looking more or less like the curve illustrated in Figure 3. This type of curvature (concave up) can arise in many ways. If the receptor population is heterogeneous in its affinity for the ligand, the Scatchard plot will always be concave up. Thus, a cell with two populations of receptors that bind the same ligand, one with high and one with low affinity, will exhibit curvature similar to that shown in Figure 3. For any heterogeneous receptor population binding a monovalent ligand, the initial and final slopes of the Scatchard plot, m_0 and m_∞ (see Figure 3), are (Goldstein, 1975)

$$m_0 = - < K^2 > / < K > \tag{13a}$$

$$m_\infty = -(< 1/K >)^{-1} \ . \tag{13b}$$

If there are a total of R_{iT} receptor sites that bind the ligand with affinity K_i then

(14a) $$< K > = \sum_{i=1}^{k} K_i R_{iT}/R_T$$

(14b) $$< K^2 > = \sum_{i=1}^{k} K_i^2 R_{iT}/R_T$$

(14c) $$< 1/K > = \sum_{i=1}^{k} (\frac{1}{K_i}) R_{iT}/R_T$$

where the total concentration of receptor binding sites is

(15) $$R_T = \sum_{i=1}^{k} R_{iT} \quad .$$

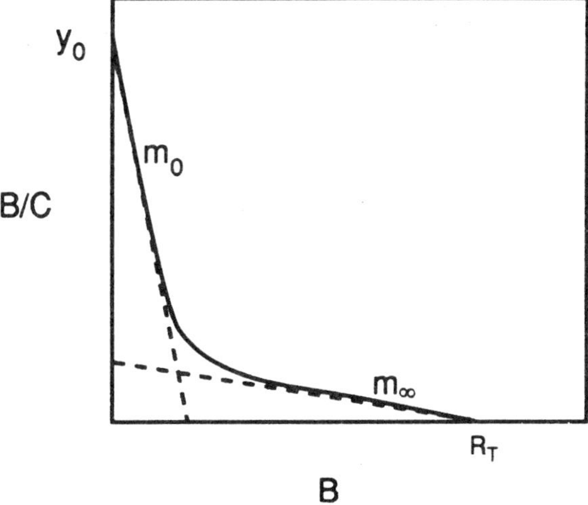

Figure 3. Scatchard plot. B and C denote the bound and free ligand concentrations, respectively; m_0 and m_∞ denote the initial and final slopes. The B intercept R_T is the total receptor site concentration.

When Scatchard plots are concave up, it is usual to fit the data to a two-receptor population model. If one of the populations has a much higher affinity than the other, so that $K_1 \gg K_2$ but the binding site concentrations for the two populations are comparable, then

(16a) $$m_0 \approx -K_1$$

(16b) $$m_\infty \approx -K_2 R_{2T}/R_T \quad .$$

Thus, if the initial slope of the Scatchard plot is much steeper than the final slope, the magnitude of the initial slope equals the equilibrium binding constant for the high affinity population, provided that the model is correct and there is more than one population of receptors on the cell surface that binds the ligand. [It follows from Eqs. (14a) and (14b) that with any number of receptor populations, if one affinity is much higher than the others, the initial slope of the Scatchard plot will have absolute value approximately equal to the highest affinity, provided that there is not a compensating imbalance in binding site densities.] It is the observation of a steep initial slope in the Scatchard plots for many of the binding studies carried out with growth factors and cytokines that has led to the view that these ligands bind their receptors with high affinity, $K = 10^{10} - 10^{11}\ \mathrm{M}^{-1}$.

For ligands like PDGF that are bivalent, there is an alternative explanation for the observed curvature in the Scatchard plot. The Scatchard plot for a bivalent ligand with two identical binding sites interacting with a homogeneous receptor population, such as depicted in Figure 1, will have the following limiting slopes (reviewed in Wofsy and Goldstein, 1992).

$$m_0 = -2K(1 + K_X R_T)^2/(1 + K_X R_T/2) \tag{17a}$$

$$m_\infty = -2K/(1 + K_X R_T/2) \ . \tag{17b}$$

When the cell surface receptor population is sufficiently high (i.e., $K_X R_T \gg 1$, which is the condition for significant dimer formation over a wide range of ligand concentrations), then

$$m_0 \approx -4K K_X R_T \tag{18a}$$

$$m_\infty \approx -4K/(K_X R_T) \ . \tag{18b}$$

For bivalent ligands to bind receptors with an apparent affinity of $10^{10}\ \mathrm{M}^{-1}$ the initial slope of the Scatchard plot $m_0 \approx -4K K_X R_T = -10^{10}\ \mathrm{M}^{-1}$. If, for example, the single site binding constant of the receptor, K, is $10^8\ \mathrm{M}^{-1}$, then to achieve an apparent affinity of $10^{10}\ \mathrm{M}^{-1}$, one must have $4K_X R_T \approx 10^2$. Can this be achieved under physiological conditions?

2.3. Estimating Equilibrium Crosslinking Constants for Dimer Formation. If we know the value of the single site equilibrium constant K for the binding of a site on a bivalent ligand in solution to a cell surface receptor, how can we estimate the value of the equilibrium crosslinking constant K_X for dimer formation; i.e., how is the three-dimensional equilibrium binding constant K related to the two-dimensional crosslinking constant K_X? Crothers and Metzger (1972) addressed this question when considering the binding of bivalent antibodies to cell surface binding sites. The derivation we present is in the spirit of their calculation.

When one end of a bivalent ligand is bound to a receptor, the other end undergoes restricted motion until it encounters a receptor and binds to it. This free end "sees" an effective three dimensional concentration C_e such that

$$K_X R_T = K C_e \ . \tag{19}$$

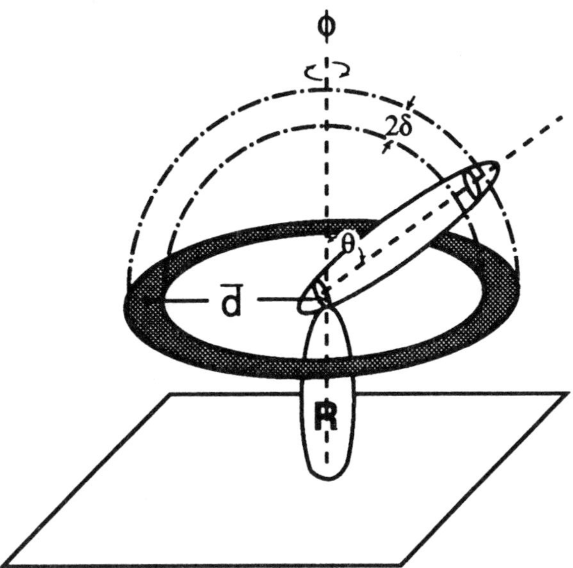

Figure 4. Hemispherical shell sampled by the free binding site on a bivalent ligand of average length d bound to a receptor R.

To estimate C_e we must specify the region in space that is accessible to the free end. In addition, for each small volume Δv_i in this region we must assign a weight w_i, which is the fraction of time the end spends in the volume Δv_i. We make the simplifying assumption that all the $w_i = 1$; i.e., the free end of the ligand is equally likely to be anywhere in the accessible region. To specify the accessible region we use a highly simplified picture of a ligand tethered to a receptor (see Figure 4). We suppose that the distance between binding sites on the ligand is d and that this distance fluctuates about its average $\bar{d}$ so that the end is uniformly distributed between $\bar{d} - \delta$ and $\bar{d} + \delta$. Because the receptor is free to rotate about its axis, there is no rotational restriction about the z-axis, and $0 \leq \phi \leq 2\pi$. We know nothing about the allowable range for θ except that angles near $\theta = \pi/2$ are accessible because the ligand can aggregate receptors into dimers. We assume that the end of the ligand that is bound to the receptor is free to rotate so that the ligand sweeps out a hemisphere, i.e., $0 \leq \theta \leq \pi/2$. For this model, shown in Figure 4, the accessible region is a hemispherical shell of thickness 2δ with approximate volume $4\pi\delta\bar{d}^2$. The number of cell surface receptors within this volume = (the number of receptors per unit area) × (the intersection of the surface area with the accessible volume) = $R_T \times (4\pi\bar{d}\delta)$. Thus $C_e = R_T/\bar{d}$, and from Eq. (19) we have that

$$K_X = K/\bar{d} \quad . \tag{20}$$

If the tethered ligand is relatively flexible, the details of how one estimates the accessible volume have little effect on Eq. (20). For example, if we assume the free end of the ligand is equally likely to be anywhere in a hemisphere of radius $\bar{d}$ rather then in a small shell, we find that $K_X = 1.5K/\bar{d}$; if we assume it can be anywhere in a sphere of radius $\bar{d}$, then $K_X = 0.75K/\bar{d}$; while if we assume it can be anywhere in a spherical shell rather than a hemispherical shell of thickness 2δ, then $K_X = 0.5K/\bar{d}$. [One could multiply the right side of Eq. (20) by a factor $\alpha \approx 0.5 - 1.5$, whose value differs from ligand to ligand and is determined by the ligand's flexibility when one of its ends is receptor bound, but our interest is in using Eq. (20) for order of magnitude estimates.]

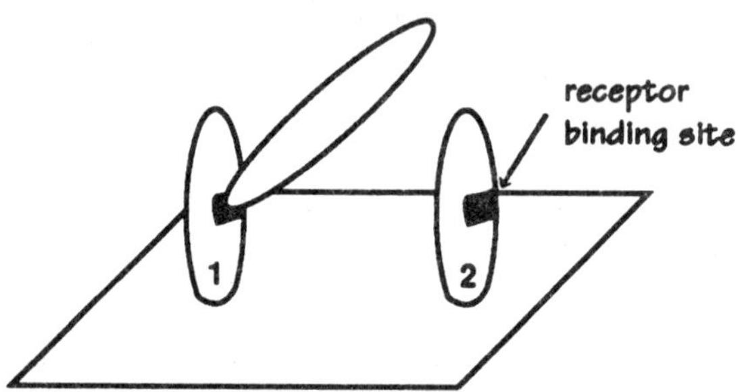

Figure 5. Steric hindrance of crosslink formation.

In deriving Eq. (20) we have ignored the size of the receptors. Since a receptor is not a point-binding site but a three-dimensional object, it may need to be correctly oriented (as well as in the accessible volume) for it to be bound by the free end of the ligand. This is illustrated in Figure 5. (In this picture, because receptors are free to rotate and the ligand is long compared to the receptor, receptor 2 will spend roughly half its time in a configuration that can bind to the ligand.) By ignoring steric hindrance we have overestimated the concentration of accessible sites and therefore overestimated C_e. Steric hindrance becomes less important as the ligand becomes more flexible (less like a rigid rod). Because flexibility increases with ligand length (the longer a molecule the more bond angles it has and the more flexible it becomes) the effects of steric hindrance become negligible for long ligands. Thus, one might modify Eq. (20) by introducing a steric hindrance function $H(\bar{d})$ so that

$$K_X = KH(\bar{d})/\bar{d} \tag{21}$$

where in the limit that $\bar{d} \to \infty$, $H(\bar{d}) = 1$, and for $\bar{d}$ below some cut-off value, the distance of closest approach of two binding sites, $H(\bar{d}) = 0$.

Dembo and Goldstein (1978a,b) have presented a model for calculating steric hindrance effects for bivalent ligands binding to antibodies in solution and on cell surfaces. Since, with the exception of the human growth hormone (hGH) receptor (de Vos et al., 1992), much less is known about the structure of growth factor and cytokine receptors than antibodies, we shall not attempt this.

If there are N receptors on a cell with surface area A, then from Eqs. (19) and (21) we have

$$H = K_X R_T \leq \frac{NK}{\bar{d}A} \tag{22}$$

where the equality corresponds to $H = 1$. Using typical values, $\bar{d} = 4 \times 10^{-7}$ cm, $A = 5 \times 10^{-6}$ cm^2, $N = 10^4$ receptors/cell, and $K = 10^8$ M^{-1}, Eq. (22) predicts that $K_X R_T \leq 10^3$. If $K_X R_T = 10^3$ and $K = 10^8$ M^{-1}, a Scatchard plot would have an initial slope $m_0 = -4 \times 10^{11}$ M^{-1}. Thus, if steric hindrance is not dominating, a bivalent ligand can exploit the effective high concentration that confining receptors to a surface creates, and be bound at concentrations much lower than its single site equilibrium constant would allow.

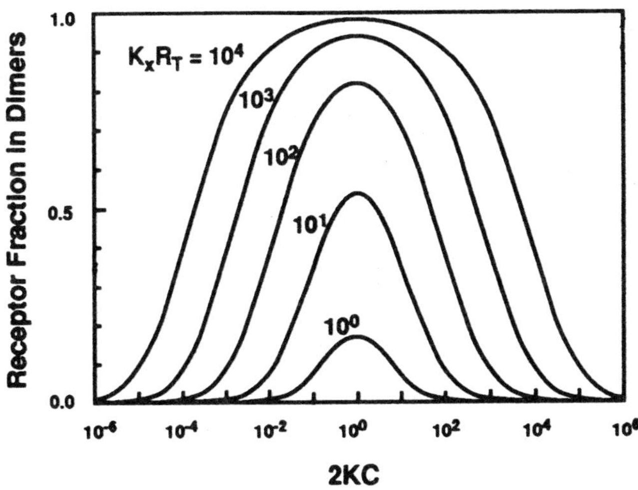

Figure 6. Predicted fraction of receptors in dimers, as a function of the ligand concentration, for different values of the lumped parameter $K_X R_T$ that characterizes the tendency for crosslinks to form. The range of ligand concentrations for which dimer formation is near maximal increases with $K_X R_T$.

2.4. High Dose Inhibition. A characteristic of a process activated through the formation of receptor dimers bridged by multivalent ligands (such as PDGF induced tyrosine phosphorylation) is that the process falls off at high ligand concentration; i.e., it exhibits high dose inhibition (see Figure 2). As the ligand concentration goes to infinity, all receptors become bound monovalently and the concentration of receptor dimers goes to zero. This raises the question of what is meant by a high ligand concentration; i.e., how high must one make the ligand concentration to see a decrease in the process being monitored? Figure 6 shows the predicted concentration of dimers as a function of the ligand concentration for different values of $K_X R_T$. As Figure 6 indicates, for there to be significant dimer formation over a wide range of ligand concentrations, $K_X R_T = HKN/(\bar{d}A) \gg 1$. If the process of interest requires only a small number of receptor dimers to produce a maximal response, the response can be maximal over a wide range of ligand concentrations. For example, if $K_X R_T = 10^2$, and if 10% of the receptors in dimers produces a maximal response, the response will be maximal over approximately six orders of magnitude, something that might make high dose inhibition hard to detect. Recently, high dose inhibition has been seen for a hybrid receptor containing the extracellular domain of human growth hormone (hGH) linked to the transmembrane and intracellular portion of murine granulocyte colony-stimulating factor (mG-CSF) (Fuh et al., 1992). The response that was observed, radioactively labeled thymidine incorporation (a measure of cell proliferation), was maximal over approximately four orders of magnitude.

2.5. Which Cytokines are Bivalent? The growth factor we have considered, PDGF-BB, as well as PDGF-AA, is composed of two identical subunits connected by disulfide bonds, while PDGF-AB is a heterodimer. Interferon-γ (Ealick et al., 1991), macrophage colony-stimulating factor (Pandit et al., 1992), nerve growth factor (McDonald et al., 1991), and stem cell factor (Zsebo et al., 1990) are homodimers. Their dimeric structure suggests, but does not guarantee, that they act as bivalent ligands, capable of bridging two receptors. Steric hindrance or allosteric changes induced when the first site binds might prevent a ligand from binding two identical receptors simultaneously even though the ligand itself is composed of two identical subunits. This seems to be the case for human interleukin-5 (IL-5) , which is a disulfide bonded homodimer (Sanderson, 1990) but does not dimerize soluble human IL-5 receptor (Devos et al., 1993).

The crystal structure of hGH, complexed with the extracelluar domain of its receptor, directly demonstrated that hGH bridges two hGH receptors (de Vos et al., 1992). However, hGH does not act as a simple bivalent ligand with two noninteracting binding sites. The two binding sites of hGH are not identical; site 1 has high affinity ($K \approx 10^9\ M^{-1}$) while site 2 has low affinity ($\leq 5 \times 10^6\ M^{-1}$); in the absence of site 1, binding of site 2 to its receptor was not detected. When site 1 binds to its receptor, site 2 is converted to a high affinity site (Cunningham et al., 1991). Undoubtedly other ligands, yet to be elucidated, will also use this cooperative mechanism to induce receptor aggregation.

2.6. Some Monovalent Ligands May Also Induce Receptor Aggregation. The binding of epidermal growth factor (EGF) to its receptors causes these receptors to dimerize rapidly (Yarden and Schlessinger, 1987; Böni-Schnetzler and Pilch, 1987; Cochet et al., 1988) despite EGF being monovalent (Weber et al., 1984). Even in the absence of the ligand, small amounts of EGF receptors aggregate in dimers (Cochet et al., 1988). It appears as if EGF receptors are somewhat "sticky" for each other, and binding of EGF enhances this "stickiness."

The binding and aggregation reactions that are thought to occur when a monovalent ligand such as EGF induces receptors to dimerize are illlustrated in Figure 7. From detailed balance we see that $KK_{X1} = K_1 K_X$, or equivalently

$$K_{X1}R_T = K_X R_T (K_1/K) \quad . \tag{23}$$

If receptor dimers initiate a biological signal, and the system is responsive to small concentrations of the ligand, then we expect binding at low ligand concentrations to induce receptor dimerization. At equilibrium a significant fraction of the bound receptors will be in dimers (composed of a bound and unbound receptor) if $K_{X1}R_T \geq 1$. In the absence of the ligand we expect the fraction of receptor dimers to be small, as in the case of the EGF receptor (Cochet et al., 1988). If we assume that 1% of the receptors are in dimers in the absence of the ligand, one can show that $K_X R_T \approx 0.01$ (Wofsy et al., 1992). Therefore $(K_1/K) \geq 10^2$. If a monovalent ligand induces receptor aggregation, the equilibrium binding constant for binding to a free site in a dimer with both sites free must be higher than for binding to a free site on a receptor that is not in a dimer.

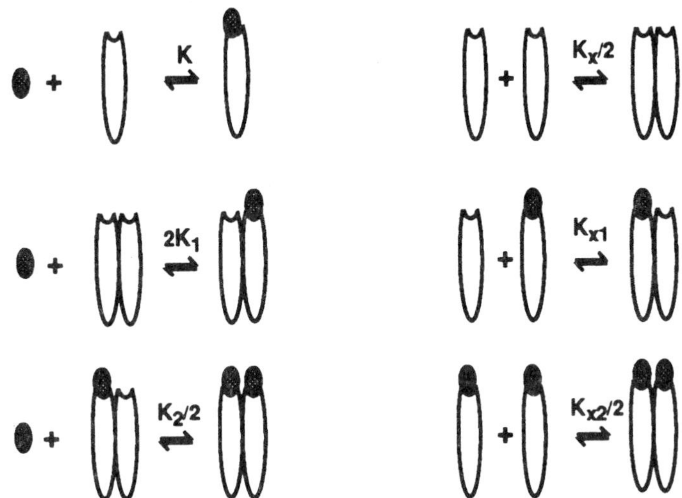

Figure 7. Binding and aggregation reactions when a monovalent ligand such as EGF induces receptors to dimerize.

In some receptor systems the binding of a ligand to a receptor induces that receptor, possibly through an allosteric change, to aggregate with one or more additional components. For example human IL-3, IL-5, and granulocyte-macrophage colony stimulating factor, GM-CSF, all bind to different α chains (the receptors) that aggregate with the same β chain (the common signaling unit) (Miyajima et al., 1992). In the absence of β the binding is low affinity; in the absence of α there is no detectable binding; but in the presence of both chains the binding is high affinity. The ratio of these affinities is typically between 10 and 10^3.

In summary, cytokines induce receptor aggregation in different ways, by being multivalent, by allosteric mechanisms, and possibly even by a combination of both (Lev et al., 1992). Many of these receptor systems await detailed models that can be used to analyze experiments rigorously and to test ideas about how dimerization comes about.

3. Antigens and the Formation of Large Aggregates

The largest aggregate that can form when a monovalent receptor interacts with a multivalent ligand is a complex of one ligand and ν receptors, where ν is the valence of the ligand (e.g., a trivalent ligand can bridge at most three monovalent receptors.) However, if a receptor has two or more binding sites for a ligand that is also multivalent, the possibility arises that very large aggregates can form. Antibodies have valences of two or greater and they bind to mutivalent antigens. When they are mixed in the appropriate concentrations, large aggregates are frequently observed. Indeed, this property of antibody-antigen reactions is used to initiate precipitation reactions. When cell surface antibodies are fluorescently labeled, exposure to antigen often leads to a redistribution of the fluorescence, from uniform to patchy, indicating that large antibody-antigen aggregates have formed.

Antibodies that act as cell surface receptors for antigens can be either noncovalently associated with integral membrane proteins (Fc receptors) or integral membrane proteins themselves (surface immunoglobulin). As surface immunoglobulin on B cells, they span the membrane and participate in large multi-subunit antigen receptor complexes (Clark et al., 1992). As solution antibodies they become transiently associated with cells of the immune system, including B cells, that express Fc receptors (Ravetch and Kinet, 1991). Fc receptors bind to the constant region (the non-antigen binding region) of specific classes of antibodies in a way that probably does not alter, and certainly does not block, the antibodies' ability to bind antigen. While an antibody is bound by an Fc receptor to the surface of a cell, it acts as a receptor for antigen. The lifetime of an antibody-Fc receptor complex varies widely. In the absence of antigen it can be a few seconds or many hours, depending on the particular Fc receptor involved.

The Fc receptor with highest affinity for antibody ($K \approx 10^{10}$ M^{-1}), now known as $\mathrm{Fc}_\epsilon\mathrm{RI}$, is specific for immunoglobulin E (IgE). This receptor is found on basophils and mast cells and plays a central role in allergic reactions of the immediate hypersensitive type. Once bound to surface $\mathrm{Fc}_\epsilon\mathrm{RI}$, IgE dissociates slowly, with a half-life of hours (Kulczycki and Metzger, 1974). For many experiments this long half-life allows one to neglect IgE dissociation and treat IgE as if it were an integral membrane protein. In the theory presented below we make this approximation.

$\mathrm{Fc}_\epsilon\mathrm{RI}$ is composed of four chains, an α chain that is responsible for binding IgE, a β chain whose function is unknown, and two disulfide-linked γ chains that participate in cell signaling events (Metzger, 1991). Although none of the chains has intrinsic kinase domains, once these receptors are aggregated, rapid tyrosine phosphorylation is observed on the β and γ chains of the receptor and on other distinct proteins (Paolini et al., 1991; 1992; Benhamou and Siraganian, 1992; Li et al., 1992; Pribluda and Metzger, 1992). To initiate any signal via the $Fc_\epsilon RI$ receptor requires the aggregation of $Fc_\epsilon RI$ receptors.

The simplest ligand that can induce IgE aggregation is one with two binding sites. We will restrict our discussion to these bivalent ligands. (Theory has been presented for ligands with valence greater than two (e.g., Goldstein and Perelson, 1984; Macken and Perelson, 1985), but binding studies with well characterized multivalent ligands have yet to be reported.) Bivalent ligands binding to bivalent receptors (such as IgE) can form only rings and chains, unlike higher valent ligands that can form networks. It is the inability to form networks that allows for considerable simplification in the modeling of bivalent ligand-bivalent receptor aggregation.

1. Sensitization of RBL Cells with FITC-anti-DNP IgE

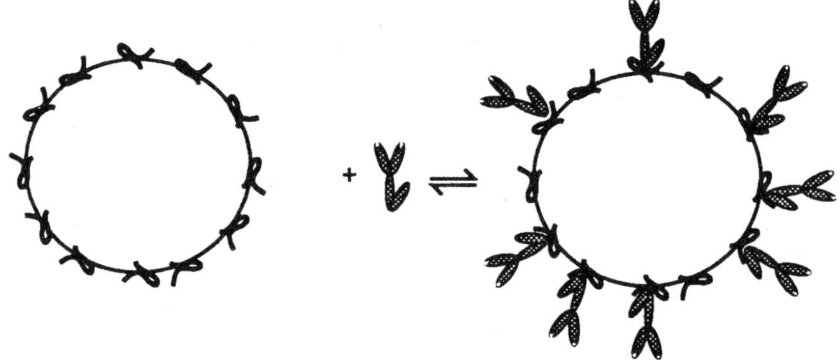

2. Exposure of sensitized RBL cells to a bivalent DNP ligand

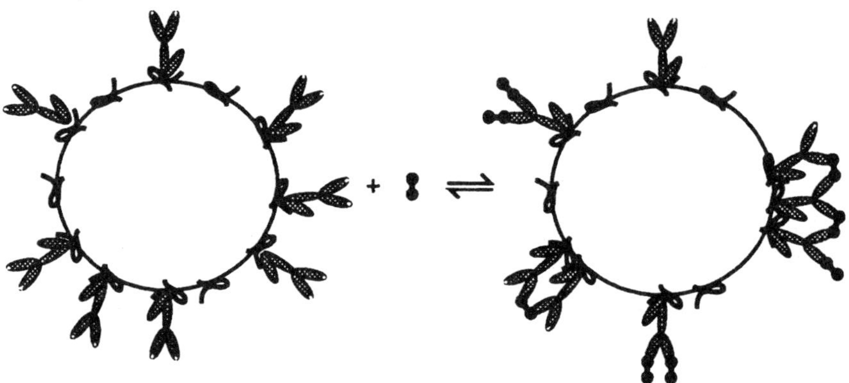

Figure 8. Experimental system. Binding of ligand to fluorescently labeled receptor (IgE) is measured by fluorescence quenching.

3.1. Bivalent Ligands Interacting With Bivalent Receptors. An experimental system introduced by Erickson et al. (1986; 1991a) that is ideally suited for the study of receptor aggregation consists of the following: rat basophilic leukemia (RBL) cells, which have on their surface $2-6\times 10^5$ $Fc_\epsilon RI$ receptors; monoclonal anti-2,4-dinitrophenyl (DNP) IgE labeled with fluorescein-5-isothiocyanate (FITC); and monovalent (1 DNP group) and bivalent (2 DNP groups) ligands. RBL cells are incubated with FITC anti-DNP IgE, after which the RBL cells are said to be sensitized (to DNP or, in general, to the antigen the IgE can bind to). The system is illustrated in Figure 8. The binding of DNP ligands to cell surface or solution FITC-IgE is monitored by measuring the fluorescence quenching that accompanies DNP occupation of the FITC-IgE binding site. The ligand we focus on is a semi-rigid symmetric bivalent ligand of about 49Å, (DNP-aminocaproyl-L-tyrosyl)$_2$-cysteine, which we denote as (DNP)$_2$-cys.

Before looking at bivalent ligands interacting with bivalent receptors, we review a solution-binding study done with the bivalent ligand, (DNP)$_2$-cys, and a monovalent receptor, FITC-Fab′. With this system, at equilibrium, Erickson et al. (1986) detected binding and dimer formation and determined that $K = 2.2\times 10^9$ M^{-1} and $K_X = 3.5\times 10^7$ M^{-1} (the notation is the same as in Figure 1).

In three dimensions (3D):

$$K_X = H_{3D}K \tag{24}$$

where H_{3D} is the steric hindrance factor for bridging two receptors in solution (Dembo and Goldstein, 1978a). [Recall that when both receptors are on a surface, the equivalent relationship between K and K_X is given by Eq. (21).] As in the two-dimensional case, H_{3D} is a function of the length and flexibility of the bivalent ligand. If the ligand is shorter than the minimal distance needed to bridge two receptors, $H_{3D}=0$; while if it is long compared to the size of the receptor, and if it is flexible, $H_{3D}\approx 1$ (in this limit the binding of the two ends of the ligand become uncorrelated). We see for (DNP)$_2$-cys binding to FITC-Fab′ that there is considerable steric hindrance, $H_{3D}=0.02$, probably because the binding pocket of the Fab′ is deep and swallows up a substantial portion of the ligand.

When we study the binding of (DNP)$_2$-cys with surface IgE, there are typically 2×10^5 receptors per RBL cell, corresponding to a surface density of about 1×10^{11} binding sites/cm^2. The single site equilibrium constant K is the same for binding to IgE in solution or IgE anchored to an Fc receptor (Erickson et al., 1986), so that $K_X R_T = H_{2D} 7.4\times 10^5$. [$H_{2D}$ is the two-dimensional steric hindrance factor that was introduced in Eq. (21). We have added a subscript to distinguish it from the three-dimensional steric hindrance factor in Eq. (24).] When $K_X R_T \geq 1$, there will be significant receptor aggregation. Therefore, we expect (DNP)$_2$-cys to be a good ligand for aggregating surface IgE, provided that $H_{2D} > 1\times 10^{-6}$, a number that is four orders smaller than H_{3D}. Not surprisingly, (DNP)$_2$-cys induces the aggregation of FITC-IgE on RBL cells (Erickson et al., 1986; 1991b).

Even though $(DNP)_2$-cys binds and aggregates IgE, it is a poor initiator of cell signaling events. Although receptor aggregation is necessary for triggering an IgE mediated RBL response, it is not sufficient. What the requirements are for an IgE aggregate to be a viable signaling unit is a major unanswered question.

As a first step toward answering this question we would like to know whether our bivalent ligand forms large or small aggregates with IgE on RBL cells. More specifically, we would like to be able to predict what the distribution of aggregates is on the RBL cell surface as a function of time after the addition of $(DNP)_2$-cys. Although as yet we have not completely solved this problem, we will outline the approach we are using.

3.2. The Kinetics of Bivalent Ligand-Bivalent Receptor Aggregation: The Equivalent Site Approximation. At first sight the problem of describing mathematically the kinetics of binding of a bivalent ligand to a bivalent receptor seems formidable. If one wants to calculate how the concentrations of aggregates change in time, then it would seem that one has to write down a chemical rate equation for each aggregate. This would result in an infinite system of coupled ordinary differential equations. However, there is a trick that reduces the number of differential equations to two. It is based on the equivalent site approximation. Perelson and DeLisi (1980) were the first to use this approximation for bivalent receptor-bivalent ligand binding, and we will follow their formulation. If rings form, the equivalent site approximation will have to be modified, but for the moment let us assume rings do not form.

The equivalent site approximation makes the following assumption: all free receptor sites, no matter what size aggregate they are in, have identical binding properties. So, for example, the single site rate constant for a ligand with both sites free binding to a free site on an isolated receptor is the same as the single site rate constant for a ligand with both sites free binding to a free site on a receptor at the end of a chain of crosslinked receptors, and this is true for all size chains. Similarly, if the equivalent site approximation is valid, the rate constants for singly bound ligands to crosslink free receptor sites are the same, no matter what size chains the ligand and free receptor sites are in.

The trick that Perelson and DeLisi used was to keep track of the states of the ligands rather than of the aggregates. This can be done because in the equivalent site approximation we can fully describe a ligand by indicating whether it has both sites free, one free and one bound, or both bound. If receptor sites were not equivalent, then to fully describe a bound ligand one would also have to specify what size aggregate it was bound to, and this would lead to an infinite set of differential equations.

The second important feature of the equivalent site approximation is that one can go from a description of ligands to a description of aggregates. That is, Perelson and DeLisi (1980) showed that if one knows how the concentrations of free ligands, singly bound ligands, and doubly bound ligands change with time, one can calculate how the concentration of any size aggregate changes in time.

In Figure 9 we define (a) the single site rate constants, k_{+1} and k_{-1}, for a ligand with both sites free binding to a free receptor site; and (b) the single site rate constants, k_{+2} and k_{-2}, for a ligand with one site bound and one site free, binding to a free receptor site. As before, we shall call the equilibrium constants $K = k_{+1}/k_{-1}$ and $K_X = k_{+2}/k_{-2}$. Also we let S denote the concentration of free receptor binding sites, S_T the total concentration of receptor binding sites [$S_T = 2\times$ (IgE concentration)], C the concentration of ligands (not sites) with both sites free, C_T the total concentration of ligands, y_1 the concentration of ligands with one site free and one site bound, and y_2 the concentration of ligands with both sites bound. Then we have the following rate equations for singly and doubly bound ligands when there is no ring formation:

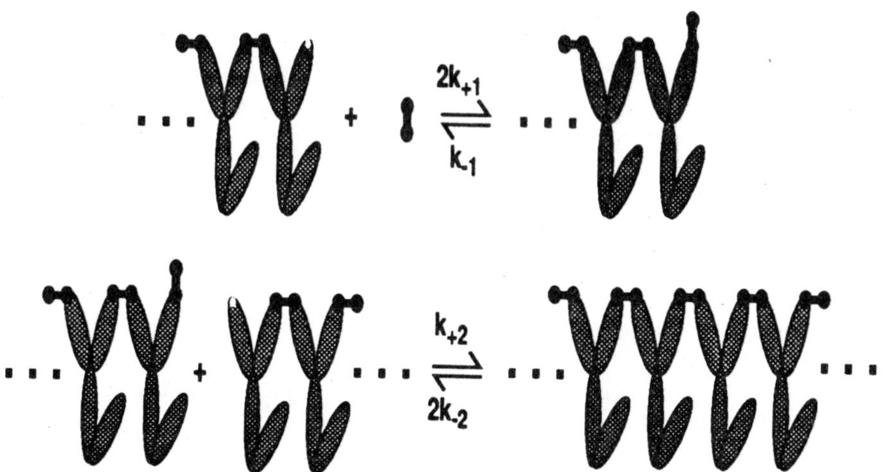

Figure 9. Reactions of a bivalent ligand with a bivalent receptor. In the equivalent site approximation there are four single site rate constants, i.e., forward and reverse rates of binding and crosslinking.

$$dy_1/dt = 2k_{+1}CS - k_{-1}y_1 - k_{+2}Sy_1 + 2k_{-2}y_2 \tag{25}$$

$$dy_2/dt = k_{+2}Sy_1 - 2k_{-2}y_2 \ . \tag{26}$$

We also have the following conservation laws for total receptor sites and total ligand:

$$S_T = S + y_1 + 2y_2 \tag{27}$$

$$C_T = C + y_1 + y_2 \ . \tag{28}$$

which we can use to eliminate S and C in Eqs. (25) and (26). In the equivalent site approximation, in the absence of ring formation, the two differential equations and two conservation laws give a complete description of the formation and dissociation of chains of all lengths.

By solving Eqs. (25) - (28) numerically, one can predict how the concentration of free ligands, singly bound ligands, doubly bound ligands and free receptor binding sites, i.e., C, y_1, y_2, and S, change with time. If the equivalent site approximation is valid, knowing this is sufficient to predict the time course of any size aggregate. For example, consider the aggregate in Figure 10 that is composed of one free site, two doubly bound ligands and one singly bound ligand. If all sites are equivalent we expect its concentration to be proportional to $Sy_2^2y_1$. Perelson and DeLisi (1980) showed that in the absence of rings the concentration of an aggregate containing n receptors and j singly bound ligands is

$$c_j(n) = \binom{2}{j} \frac{S_T}{2} \left(\frac{y_1}{S_T}\right)^j \left(\frac{2y_2}{S_T}\right)^{n-1} \left(\frac{S}{S_T}\right)^{2-j} \quad . \tag{29}$$

Since singly bound ligands can occur only on the ends of chains, $j = 0$, 1 or 2. For the aggregate in Figure 10, $n = 3$ and $j = 1$, so its concentration $c_1(3) = y_1(2y_2/S_T)^2(S/S_T)$. The concentration of any other aggregate can be obtained in a similar way.

Figure 10. An aggregate with one free receptor binding site, two doubly bound ligands and one singly bound ligand.

If rings are present, sites in rings are not equivalent to sites in chains. In fact if all sites were equivalent, rings would never form. For a chain, such as the one shown in Figure 10, to form a ring, the free site on the singly bound ligand must bind to the free site on the end of the chain rather than to any of the other free sites in the population. This selective binding event would have probability zero if all sites were equivalent. In any modification of the equivalent site approximation to include ring formation, sites in rings of different sizes must be handled separately. First, ring closure rates are unlikely to be the same for different sized rings. Second, when a ring with n doubly bound ligands opens, n doubly bound ligands in rings are converted to one singly bound ligand and $n-1$ doubly bound ligands in chains. This is not true for chains. When a chain opens, one doubly bound ligand is converted to one singly bound ligand, independent of the size of the chain the doubly bound ligand was in. Thus if rings of all sizes form, we are back to the problem of having to solve an infinite system of differential equations.

Since the effect of ring formation is to terminate chains, if rings form the concentration of long chains will be reduced. This in turn means that the concentration of large rings will be small. As a simple first approximation it seems reasonable to assume that there is only one or a few sizes of rings that can have significant concentrations. The bivalent ligand we have studied, $(DNP)_2$-cys, is too small to bridge both bindings sites on the same IgE. The smallest ring it can form with IgE is of size 2, i.e., two doubly bound ligands and two IgEs. If this were the only ring that could form, we would have to keep track of just one additional ligand state, doubly bound ligands in rings of size 2.

If we call the concentration of chains made up of two crosslinked IgEs with a bivalent ligand bound to one end of the chain $c_1(2)$, the chemical rate equation describing the formation and dissociation of ligands in rings of size two is

$$dr_2/dt = 2j_{+2}c_1(2) - 4j_{-2}r_2 \tag{30}$$

where r_2 is the concentration of doubly bound ligands in rings of size two, and j_{+2} and j_{-2} are the rate constants for the closing and opening of a ring of size 2 (see Figure 11). (Note that when you form a ring, you gain two ligands that are in a ring of size two. That is why the first term on the right side of Eq. (31) is multiplied by two. The last term is multiplied by four because a ring of size two can dissociate if any one of four sites opens.) In the presence of these rings Eqs. (25) and (26) are modified and become

$$dy_1/dt = 2k_{+1}CS - k_{-1}y_1 - k_{+2}Sy_{-1} + 2k_{-1}y_2 - j_{+2}c_1(2) + 2j_{-2}r_2 \tag{31}$$

$$dy_2/dt = k_{+2}Sy_1 - 2k_{-2}y_2 - j_{+2}c_1(2) + 2j_{-2}r_2 \ . \tag{32}$$

When rings of size two are included, the conservation laws become

$$S_T = S + y_1 + 2y_2 + 2r_2 \tag{33}$$

$$C_T = C + y_1 + y_2 + r_2 \ . \tag{34}$$

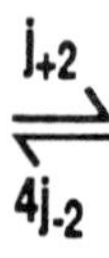

Figure 11. Formation and break-up of a ring consisting of two antibodies and two bivalent ligands.

Perelson and DeLisi (1980) have written down the chemical rate equations for the general case when any ring can form. However, in their formulation they used Eq. (29) to express $c_j(n)$ in terms of y_1, y_2, and S, which is incorrect if rings form (Posner, unpublished result). Elsewhere we show how to modify the equivalent site approximation to treat the formation of rings of a finite set of sizes (Posner et al., 1994). Here we simply describe one experiment where the knowledge of the correct form of $c_j(n)$ is not required.

Consider the following dissociation experiment. RBL cells sensitized with fluorescently labeled anti-DNP are exposed to a bivalent ligand, $(\mathrm{DNP})_2$-cys. After a time t_1, a large excess of monovalent ligand is added, rapidly filling all free sites. Further, as sites bound by bivalent ligands open, they are immediately filled by monovalent ligand so that the concentration of free sites, S, is always zero. This means that after the addition of monovalent ligand, rings that open can never re-form (for $t \geq t_1$, $S = 0$ and $c_1(n) = 0$ for all n). Thus, for $t \geq t_1$

$$dy_1/dt = -k_{-1}y_1 + 2k_{-1}y_2 + 2j_{-2}r_2 \tag{35}$$

$$dy_2/dt = -2k_{-2}y_2 + 2j_{-2}r_2 \tag{36}$$

$$dr_2/dt = -4j_{-2}r_2 \tag{37}$$

where at $t = t_1$, $y_1 = y_1(t_1)$, $y_2 = y_2(t_1)$ and $r_2 = r_2(t_1)$. It is straightforward to solve these equations and show that the concentration of bound sites, $S_b = y_1 + 2y_2 + 2r_2$, for $t \geq t_1$ is (Posner et al., 1991)

$$S_b = A_1e^{-k_1t} + A_2e^{-2k_2t} + A_3e^{-4j_2t} \tag{38}$$

where, if $k_1 = k_2$, $A_2 = 0$ and if $k_2 = j_2$, $A_3 = 0$. Because the amount of fluorescence quenching per DNP group is different for the monomer and dimer, the time course of S_b can be monitored after the addition of monomer. We find that the recovery of the fluorescence is well fit by two exponentials with exponents of approximately 5×10^{-2} s^{-1} and 3×10^{-3} s^{-1} (Erickson et al., 1991b). Our working hypothesis is that the bivalent ligand tends to form rings rather than long chains. [The tendency for small bivalent ligands to form rings has previously been observed by Schweitzer-Stenner et al. (1987)]. Bound sites on singly and doubly bound ligands in chains open at the same rate ($k_1 = k_2 = 3 \times 10^{-3}$ s^{-1}) while the faster decay corresponds to the opening of rings ($4j_2 = 5 \times 10^{-2}$ s^{-1}). Why small rings made of two IgEs and two ligands are poor at triggering cellular responses is unclear, but it is known that larger aggregates are more effective in stimulating RBL cells to degranulate than small ones (Fewtrell and Metzger, 1980.)

In summary, the equivalent site approximation as introduced by Perelson and DeLisi (1980) is an elegant way to describe the kinetics of bivalent ligand-bivalent receptor binding and aggregation. The approach works for small bivalent ligands interacting with solution or surface antibody, provided that only chains of crosslinked receptors form. However, experiments indicate that small bivalent ligands tend to aggregate antibodies into small rings (Schweitzer-Stenner et al, 1987; Erickson et al., 1991b). In addition, receptor aggregates can interact with other cellular components that dramatically reduce receptor mobility. Both ring formation and altered mobility of receptor aggregates mean that the equivalent site approximation must be modified to describe the time course of aggregate formation on cell surfaces.

4. Conclusion

Ligand induced receptor aggregation is a ubiquitous cell signal. For a wide variety of ligands, binding is insufficient to initiate a cellular response, rather the aggregation of specific cell surface proteins is required. Although aggregation is necessary, it is not sufficient. Some antibodies that bridge receptors can be substituted for the natural ligand and induce robust responses, while others, which also bridge receptors, are ineffective (Fuh et al., 1992). What the properties are that make a receptor aggregate a viable signaling unit are, to a large extent, unknown.

The interaction of cytokines with their receptors leads to the formation of small aggregates of cell surface proteins. For example, many growth factor receptors form homodimers (two identical growth factor receptors) on exposure to specific growth factor. IL-3, IL-5, and GM-CSF induce the formation of heterodimers composed of a receptor (α chain) and one common signaling unit (β chains). GM-CSF may also induce the formation of heterotrimers composed of an α and two β chains (Budel et al., 1993). IL-6 complexed with an α and two β chains has been detected (Murakami et al., 1993). IL-2 forms an aggregate with three different transmembrane proteins (Taniguchi and Minami, 1993). Antigens binding to cell surface immunoglobulin and immune complexes binding to Fc receptors often induce formation of large aggregates. The aggregates that different ligands can induce to form are varied both in size and structure.

Despite the recognition that ligands induce receptor aggregation, the models used by experimental biologists to analyze their data quantitatively often ignore aggregation. (Usually it is assumed that there are two fixed populations of receptors, one with high affinity and one with low.) There is clearly a need for mathematical models of aggregating receptor systems. These models must be custom designed for the specific ligand-receptor system of interest and the specific type of experiments to be analyzed. We expect that with close interaction between model and experiment, such models will become useful tools in helping to understand the earliest events in the initiation of cell signaling.

ACKNOWLEDGEMENTS

This manuscript was written while we were on sabbatical leave at the National Institutes of Health. We wish to acknowledge the hospitality of Dr. Henry Metzger (NIAMS) and Dr. John Rinzel (NIDDK).

REFERENCES

Bellot, F., G. Crumley, J.M. Kaplow, J. Schlessinger, M. Jaye, and C.A. Dionne, Ligand-induced transphosphorylation between different FGF receptors, *EMBO J.* **10** (1991), 2849-2854.

Benhamou, M. and R.P. Siraganian, Protein-tyrosine phosphorylation: An essential component of $Fc_\epsilon R1$ signaling, *Immunol. Today* **13** (1992), 195-197.

Böni-Schnetzler, M. and P.F. Pilch, Mechanism of epidermal growth factor receptor autophosphorylation and high-affinity binding, *Proc. Natl. Acad. Sci. USA* **266** (1987), 7839-7836.

Budel, L.M., H. Hoogerbrugge, K. Pouwels, C. van Buitenen, R. Delwel, B. Löwenberg and I.P. Touw, Granulocyte-macrophage colony-stimulating factor receptors alter their binding characteristics during myeloid maturation through up-regulation of the affinity converting β subunit (KH97), *J. Biol. Chem.* **268** (1993), 10154-10159.

Clark, M.R., K.S. Campbell, A. Kazlauskas, S.A. Johnson, M. Hertz, T.A. Potter, C. Pleiman and J.C. Cambier, The B cell antigen receptor complex: Association of Ig-α and Ig-β with distinct cytoplasmic effectors, *Science* **258** (1993), 123-126.

Cochet, C., O. Kashles, E.M. Chambaz, I. Borrello, C.R. King and J. Schlessinger, Demonstration of epidermal growth factor-induced receptor dimerization in living cells using a chemical covalent cross-linking agent, *J. Biol. Chem.* **263** (1988), 3290-3295.

Crothers, D.M. and H. Metzger, The influence of polyvalence on the binding properties of antibodies. *Immunochem.* **11** (1972), 341-357.

Cunningham, B.C., M. Ultsch, A.M. de Vos, M.G. Mulkerring, K.R. Clauser and J.A. Wells, Dimerization of the extracellular domain of the human growth hormone receptor by a single hormone molecule, *Science* **254** (1991), 821-825.

Dembo, M. and B. Goldstein, A thermodynamic model of binding of flexible bivalent haptens to antibody, *Immunochem.* **15** (1978a), 307-313.

Dembo, M. and B. Goldstein, Theory of equilibrium binding of symmetric bivalent haptens to cell surface antibody: Application to histamine release from basophils, *J. Immunol.* **121** (1978b), 345-353.

de Vos, A.M., M. Ultsch and A. Kossiakoff, Human growth hormone and extracellular domain of its receptor: Crystal structure of the complex, *Science* **255** (1992), 306-312.

Devos, R., Y. Guisez, S. Cornelis, A. Verhee, J. Van der Heyden, M. Manneberg, H.-W. Lahm, W. Fiers, J. Tavernier, and G. Plaetinck, Recombinant soluble human interleukin-5 (IL-5) receptor molecules: Cross-linking and stoichiometry of binding to IL-5, *J. Biol. Chem.* **268** (1993), 6581-6587.

Ealick, S.E., W.J. Cook, S. Vijay-Kumar, M. Carson, T.L. Nagabhushan, P.P. Trotta, and C.E. Bugg, Three-dimensional structure of recombinant human interferon-γ, *Science* **252** (1991), 698-702.

Erickson, J.W., R. Kane, B. Goldstein, D. Holowka, and B. Baird, Cross-linking of IgE-receptor complexes at the cell surface: A fluorescence method for studying the binding of monovalent and bivalent haptens to IgE, *Mol. Immunol.* **23** (1986), 769-781.

Erickson, J.W. , R. Posner, B. Goldstein, D. Holowka, and B. Baird,Analysis of ligand binding and cross-linking of receptors in solution and on cell surfaces: Immunoglobulin E as a model receptor. In: *Fluorescence in Biochemistry and Cell Biology*, G. Dewey, ed., Plenum Publishing, New York, 1991a, pp. 169-195.

Erickson, J.W. , R. Posner, B. Goldstein, D. Holowka, and B. Baird, Bivalent ligand dissociation kinetics from receptor-bound immunoglobulin E: Evidence for a time dependent increase in ligand rebinding at the cell surface, *Biochem.* **30** (1991b), 2357-2363.

Fewtrell, C. and H. Metzger, Larger oligomers of IgE are more effective than dimers in stimulating rat basophilic leukemia cells, *Proc. Natl. Acad. Sci. USA* **125** (1980), 701-710.

Fuh, G., B.C. Cunningham, R. Fukunaga, S. Nagata, D.V. Goeddel and J.A. Wells, Rational design of potent antagonists to the human growth hormone receptor, *Science* **256** (1992), 1677-1680.

Goldstein, B., Theory of hapten binding to IgM: The question of repulsive interactions between binding sites, *Biophys. Chem.* **3** (1975), 363-367.

Goldstein, B. and A.S. Perelson, Equilibrium theory for the clustering of bivalent cell surface receptors by trivalent ligands, *Biophys. J.* **45** (1984), 1109-1123.

Hart, C.E. and D.F. Bowen-Pope, Platelet-derived growth factor receptor: Current views of the two-subunit model, *J. Invest. Dermatology* **94, Suppl.** (1990), 53S-57S.

Heldin, C.-H., A. Ernlund, C. Rorsman, and L. Rönnstrand, Dimerization of B-type platelet-derived growth factor receptors occurs after ligand binding and is closely associated with receptor kinase activation, *J. Biol. Chem.* *264* (1989), 8905-8912.

Honegger, A.M., R.M. Kris, A. Ullrich and J. Schlessinger, Evidence that autophosphorylation of solubilized receptors for epidermal growth factor is mediated by intermolecular cross-phosphorylation, *Proc. Natl. Acad. Sci. USA* **86** (1989), 925-929.

Kelly, J.D., B.A. Haldeman, F.J. Grant, M.J. Murray, R.A. Seifert, D.F. Bowen-Pope, J.A. Cooper and A. Kazlauskas, Platelet-derived growth factor (PDGF) stimulates PDGF receptor subunit dimerization and intersubunit *trans*-phosphorylation, *J. Biol. Chem.* **266** (1991), 8987-8992.

Kulczycki, A.Jr. and H. Metzger, The interaction of IgE with rat basophilic leukemia cells. II. Quantitative aspects of the binding reactions, *J. Exp. Med.* **140** (1974), 1676-1695.

Lev, S., Y. Yarden and D. Givol, Dimerization and activation of the kit receptor by monovalent and bivalent binding of the stem cell factor, *J. Biol. Chem.* **267** (1992), 15970-15977.

Li, W., G.G. Deanin, B. Margolis, J. Schlessinger, and J.M. Oliver. FcϵR1-mediated tyrosine phosphorylaton of multiple proteins, including phospholipase $C_{\gamma}1$ and the receptor $\beta_{\gamma 2}$ complex, in RBL-2H3 rat basophilic leukemia cells, *Mol. Cell. Biol.* **12** (1992), 3176-3182.

Macken, C.A. and A.S. Perelson, *Branching Processes Applied to Cell Surface Aggregation Phenomena*, Lecture Notes in Biomathematics, Vol. 58, Springer Verlag, New York, 1985.

McDonald, N.Q., R. Lapatto, J. Murray-Rust, J. Gunning, A. Wlodawer, and T.L. Blundell, New protein fold revealed by a 2.3-Å resolution crysta structure of nerve growth factor, *Nature* **354** (1991), 411-414.

Menon, A.K., D. Holowka, W.W. Webb, and B. Baird, Clustering, mobility, and triggering activity of small oligomers of immunoglobulin E on rat basophilic leukemia cells, *J. Cell Biol.* **102** (1986), 534-540.

Metzger, H., The high affinity receptor for IgE on mast cells, *Clin. and Exp. Allergy* **21** (1991), 269-279.

Miyajima, A., T. Hara, and T. Kitamura, Common subunits of cytokine receptors and the functional redundancy of cytokines, *TIBS* **17** (1992), 378-382.

Murakami, M., M. Hibi, N. Nakagawa, T. Nakagawa, K. Yasukawa, K. Yamanishi, T. Taga, and T. Kishimoto, IL-6 induced homodimerization of gp130 and associated activation of a tyrosine kinase, *Science* **260** (1993), 1808-1810.

Ohtsuka, M., M.F. Roussel, C.J. Sherr, and J.R. Downing, Ligand-induced phosphorylation of the colony-stimulating factor 1 receptor can occur through an intermolecular reaction that triggers receptor down modulation, *Mol. Cell. Biol.* **10** (1990), 1664-1671.

Pandit, J., A. Bohm, J. Jancarik, R. Halenbeck, K. Koths, and S.H. Kim, Three-dimensional structure of dimeric human recombinant macrophage colony-stimulating factor, *Science* **258** (1992), 1358-1362.

Paolini, R., M.-H. Jouvin and J.-P. Kinet, Phosphorylation and dephosphorylation of the high-affinity receptor for immunoglobulin E immediately after receptor engagement and disengagement, *Nature* **353** (1991), 855-858.

Paolini, R., R. Numerof, and J.-P. Kinet, Phophorylation/dephosphorylation of high-affinity IgE receptors: A mechanism for coupling/uncoupling a large signaling complex, *Proc. Natl. Acad. Sci. USA* **89** (1992), 10733-10737.

Pawson, T. and G.D. Gish, SH2 and SH3 domains: From structure to function, *Cell* **71** (1992), 359-362.

Perelson, A.S. and C. DeLisi, Receptor clustering on a cell surface. I. Theory of receptor cross-linking by ligands bearing two chemically identical functional groups, *Math. Biosciences* **48** (1980), 71-110.

Posner, R.G., J.W. Erickson, D. Holowka, B. Baird, and B. Goldstein, Dissociation kinetics of bivalent ligand-immunoglobulin E aggregates in solution, *Biochem.* **30** (1991), 2348-2356.

Posner, R.G., C. Wofsy, and B. Goldstein, Kinetics of bivalent ligand-bivalent receptor aggregation: Ring formation and the breakdown of the equivalent site approximation (to be published, 1994).

Pribluda, V.S. and H. Metzger, Transmembrane signaling by the high-affinity IgE receptor on membrane preparations, *Proc. Natl. Acad. Sci. USA* **89** (1992), 11446-11450.

Ravetch, J.V. and J.-P. Kinet, Fc Receptors, *Annu. Rev. Immunol.* **9** (1991), 457-492.

Rönnstrand, L., L. Terracio, L. Claesson-Welsh, C.-H Heldin, and K. Rubin, Characterization of two monoclonal antibodies reactive with the external domain of the platelet-derived growth factor receptor, *J. Biol. Chem.* **263** 1988, 10429-10435.

Sanderson, C., The biological role of interleukin 5, *Int. J. Cell Cloning* **8(1)** (1990), 147-153.

Scatchard, G., The attractions of proteins for small molecules and ions, *N.Y. Acad. Sci.* **51** (1949), 660-672.

Schweitzer-Stenner, R., A. Licht, I. Lüscher, and I. Pecht, Oligomerization and ring closure of immunoglobulin E class antibodies by divalent haptens, *Biochem.* **26** (1987), 3602-3612.

Schweitzer-Stenner, R., A. Licht and I. Pecht, Dimerization kinetics of the IgE-class antibodies by divalent haptens. II. The interactions between intact IgE and haptens, *Biophys. J.* **63** (1992), 563-568.

Severinsson, L., L. Claesson-Welsh and C.-H. Heldin, A B-type PDGF receptor lacking most of the intracellular domain escapes degradation after ligand binding, *Eur. J. Biochem.* **182** (1989), 679-686.

Spaargaren, M., L. H. K. Defize, J. Boonstra and S. W. de Laat, 1991, Antibody-induced dimmerization activates the epidermal growth factor receptor tyrosine kinase, *J. Biol. Chem.* **266** 1733-1739.

Taniguchi, T. and Y. Minami, The IL-2/IL-2 receptor system: A current overview, *Cell* **73** (1993), 5-8.

Ullrich, A. and J. Schlessinger, Signal transduction by receptors with tyrosine kinase activity, *Cell* **61** (1990), 203-212.

Weber, W., P.J. Bertics and G.N. Gill, Immunoaffinity purification of the epidermal growth factor receptor: Stoichiometry of binding and kinetics of self-phosphorylation, *J. Biol. Chem.* **259** (1984), 14631-14636.

Williams, L.T, Signal transduction by the platelet-derived growth factor receptor, *Science* **243** (1989), 1564-1570.

Wofsy, C. and B. Goldstein, Interpretation of Scatchard plots for aggregating receptor systems, *Math. Bio. Sci.* **112** (1992), 115-154.

Wofsy, C., B. Goldstein, K. Lund and H.S. Wiley, Implications of epidermal growth factor (EGF) induced EGF receptor aggregation. *Biophys. J.* **63** (1992), 98-110.

Yarden, Y. and A. Ullrich, Growth factor receptor tyrosine kinases, *Annu. Rev. Biochem.* **57** (1988), 443-478.

Zsebo, K.M., J. Wypych, I.K. McNiece, H.S. Lu, K.A. Smith, S.B. Karkare, N.C. Birkett, L.R. Williams, BV. N. Satyagal, W. Tung, R.A. Bosselman, E.A. Mendiaz, and K.E. Langley, Identification, purification, and biological characterization of hematopoietic stem cell factor from buffalo rat liver-conditioned medium, *Cell* **63** (1990), 195-201.

BYRON GOLDSTEIN
THEORETICAL BIOLOGY AND BIOPHYSICS GROUP
THEORETICAL DIVISION
LOS ALAMOS NATIONAL LABORATORY
LOS ALAMOS, NEW MEXICO 87545 U.S.A.
BXG@LANL.GOV

CARLA WOFSY
DEPARTMENT OF MATHEMATICS AND STATISTICS
UNIVERSITY OF NEW MEXICO
ALBUQUERQUE, NEW MEXICO 87131 U.S.A.
WOFSY@MATH.UNM.EDU